RADIATION

RADIATION

All You Need to Know to Stop Worrying... Or to Start

Martin D. Ecker, M.D.
Norton J. Bramesco

VINTAGE BOOKS
A Division of Random House, New York

A VINTAGE ORIGINAL, June 1981
First Edition
Copyright © 1981 by Martin D. Ecker, M.D. and Norton J. Bramesco
All rights reserved under International and Pan-American Copyright Conventions. Published in the United States by Random House, Inc., New York, and simultaneously in Canada by Random House of Canada Limited, Toronto.

Library of Congress Cataloging in Publication Data

Ecker, Martin D., 1945–
    Radiation: all you need to know to stop worrying—or to start.

    Bibliography: p.
    Includes index.
    1. Radiation—Safety measures. 2. Radiation—
Physiological effect. 3. Radiation. I. Bramesco,
Norton J., 1924– II. Title.
RA569.E35   616.9'897   80–6137
ISBN 0–394–74650–3   AACR2

Book Design: Charlotte Staub

Manufactured in the United States of America

For encouraging us and for being there when we needed them most, we dedicate this book to our wives, Judith Zolondek Ecker and Ronnie Zolondek Bramesco.

Although there can be no complete picture of the workings of nature which would be intelligible to our minds, yet we can still draw pictures to represent partial aspects of the truth . . .

—Sir James Jeans,
Physics and Philosophy

Contents

An End to Innocence   ix
Acknowledgments   xiii

PART ONE: MOSTLY SCIENCE

1. The Anatomy of an Atom   3
2. Atomic Restlessness   9
3. How Radioactivity Comes About and How It Is Measured
    20

PART TWO: LIVING WITH RADIATION

4. Radiation: It's All Around Us!   35
5. The Nuclear Industry   73
6. Mutation and Cancer: The Major Effects of Ionizing Radiation
    103
7. Even Nonionizing Radiation Can Be Hazardous   123
8. The Up Side   137

PART THREE: REDUCING THE RISKS

9.   Making the Use of Ionizing Radiation Safer   153
10. Reducing the Risks in Using Nonionizing Radiation   164
11. Safety Innovations   172

PART FOUR: THE ROLE OF GOVERNMENT

12. The Nuclear Regulatory Establishment   183
     Epilogue   196
     Glossary   199
     Selected Bibliography   207
     Index   213

An End to
Innocence

During the second week of spring in 1979, America came of nuclear age.

The unlikely backdrop for this event was a small island in the Susquehanna River near Harrisburg, Pennsylvania. Three Mile Island, as it was called, has since become synonymous with peacetime nuclear catastrophe in the same way that Hiroshima came to be associated with the destructive potential of the atom during war.

Three Mile Island was the site of a nuclear power plant operated by the Metropolitan Edison Company. This facility is one of many scattered throughout the United States. As this book goes to press, 72 reactors are already operating and 91 additional facilities are in varying stages of completion.

These are not insignificant numbers. The 72 operating reactors generate roughly 13 percent of the electricity produced in this country. If the operating reactors were forced to shut down, the result would be something of an energy nightmare. But with the specter of Three Mile Island still hovering over the American scene, the possibility of "no nukes" is a very real one.

What happened at Three Mile Island was almost a perfect scenario of what could happen. Simple mechanical devices malfunctioned, human error and faulty judgment surfaced, unforeseen circumstances wreaked havoc, computerized backup systems backed off, and failsafe mechanisms just failed. The episode turned out to be one of the most harrowing in United States history.

It started in the small hours of the morning of Wednesday, March 28, 1979, when an electronic alarm sounded and warning lights blinked on to signal a breakdown somewhere in the vastly complex system. In the control room of Unit II, immediate steps were taken by the crew. Appropriate buttons were pushed and safety mechanisms ground into action.

A pump had broken down in one of the reactor systems. Just a single pump. But for want of a nail, kingdoms can be lost.

When the system is functioning normally, the great quantities of heat produced by a nuclear chain reaction are transferred to water surrounding the reactor core. This superheated radioactive water is then channeled to a steam generator, where its heat is transferred to ordinary cold water, turning it into the steam that drives a turbine and generates electricity. What started the trouble at Three Mile Island was failure of the pump that carried ordinary water into the steam generator.

But the nuclear chain reaction continued. Heat continued to be produced, elevating the temperature of the water, which was now unable to dispel its heat. Rising pressure from this water caused the reactor to shut down automatically.

A valve opened to relieve the pressure, as designed, but it failed to close. The radioactive water was thus permitted to overflow onto the floor of the containment structure, and radiation began to seep through its walls.

Because of the overflow, water level and pressure dropped, triggering an emergency cooling system. But inadvertently or through misunderstanding, this system was shut off. Without the water or the cooling system, components involved in the nuclear chain reaction overheated and burst—perhaps even melted.

Now things went from bad to worse. A sump pump transferred the water to a nearby auxiliary building. When this building flooded, radioactive steam leaked out, spreading radiation twenty miles into the Pennsylvania countryside.

As the first tense days passed, radioactive water was dumped into the Susquehanna River, and more radioactive material escaped from the auxiliary building, raising the possibility of a mass evacuation from the surrounding area. Finally, a radioactive gas bubble formed at the top of the reactor, increasing the danger of an explosion or a reactor meltdown unless it could be vented.

While a nation too close to panic held its collective breath, the bubble was finally vented and dispelled. A monumental nuclear catastrophe had been narrowly averted. In the next several days, control over all components was reestablished, and while weeks would pass before all systems were secure, the danger ended—except perhaps for the few unfortunate Pennsylvanians who had been exposed to radiation as a result of the accident. But here, of course, only time would tell.

In the aftermath of Three Mile Island, all our old fears came surging back. We now *knew* what could happen. We had lost our nuclear virginity.

The demonstrations and calls for dismantling of the nuclear power industry were not completely unwarranted. After all, radiation out of control is scary. A strange *something* you can't feel, smell, see, or hear —and therefore can't dodge. Something that can destroy you, insidiously, years after exposure.

Unfortunately, most of the post–Three Mile Island activity was generated out of emotion rather than knowledge. Serious decisions remain to be made. In a report submitted to President Carter on October 30, 1979, a special commission appointed to investigate the accident concluded that "an accident like Three Mile Island was eventually inevitable." Yet the commission did not suggest the abandonment of nuclear power: "Our findings do not, standing alone, require the conclusion that nuclear power is inherently too dangerous to permit it to continue. . . ."

The inherent ambiguity of these statements puts the burden of concern right back on the shoulders of a public already worried about radiation absorbed from other sources—medical and dental x-rays and microwave ovens, for example. What are the effects of radiation on the human body? Are the risks worth the benefits? Should we listen to the Chicken Littles or to the Pollyannas?

Out of that quandary grew the idea for this book. We believe that decisions you may have to make for yourself and your family should be made on a foundation of fact, not misinformation or hysteria. Such information is available. But you would have to dig around in libraries, read proceedings of scientific congresses, and digest medical textbooks to get it.

We've taken all the physics, physiology and histology you need to know and distilled it into simple language. And we have put all the information together in this one volume.

In the subtitle, we promised everything you should know about radioactivity to stop worrying . . . or to start. By presenting the latest information in a clear, unbiased, and, we hope, entertaining manner, this book will help you decide which.

Acknowledgments

 We wish to express our appreciation to Ronnie Bramesco for her advice and unstinting effort. In preparing the manuscript, she truly dotted the i's and crossed the t's.

Glenna Benson helped considerably with technical research, and transcribed from dictation our often long and tedious notes.

Mitchell Sacks deserves our thanks for his advice and keen insights into what types of scientific information would be appealing to the public in these times of nuclear versus anything-but-nuclear technologies.

Susan Rebell is acknowledged for her continuous support and encouragement.

For their comprehensive technical advice and assistance, our thanks go to Drs. William Choyke and J. N. McGuire of the Department of Physics and Astronomy, the University of Pittsburgh; to Dr. Robert Hoffman of the Yale University School of Medicine; and to Mr. Samuel Sperling of the Bureau of Radiologic Health, Food and Drug Administration, U.S. Department of Health and Human Services.

The following people augmented our research with ideas and scientific source material, thus helping to assure a comprehensive work: Barbara Ecker, Miriam Ecker, Dr. Steve Zeide, Dr. Robert Lowman, Dr. Milton Zaret, Dr. Ron Ablow, Dr. Sol Zolondek, and William Zweroff.

Dr. Dorothy Peck was kind enough to read our text and furnish many constructive ideas and suggestions.

Carol Pacelli provided additional secretarial assistance when time was a problem.

More than just a legal adviser, Dennis Krieger often served as a confidant in times of need.

We are deeply grateful to Gail Winston, who went far beyond her role as editor and helped us in many ways whenever we called on her.

And we wish to acknowledge the excellent editing done by Connie Day and Nancy Inglis. Their insights and suggestions were greatly appreciated.

part one

Mostly
Science

1
The Anatomy of an Atom

"Split wood, not atoms!" We see it regularly on automobile bumper stickers. But our first step in coming to understand radioactivity *must* be to "split the atom"—to examine it and identify its parts. For radiation is produced by change in the structure of atoms.

The atom is the basic unit of all matter—animal, vegetable, and mineral. Atoms make up the human body, the air that supports life, the paper on which these words are printed, the planets, and all the stars of all the galaxies in the cosmos. But exactly what is an atom? And if it is the raw material of all matter, can the number of different types of atoms even be counted?

Today, the known different varieties of atoms number only a little over a hundred. And of these, some types are extremely rare. Some don't even exist in nature but have been created during scientific experiments. Actually, only about 10 percent of the known atoms make up almost 100 percent of all matter. And only three make up such diverse substances as coal, alcohol, wood, and oil.

The staggering variety of materials in nature is possible because this handful of different types of atoms can be arranged in an almost infinite number of configurations, or compounds. Scientists of the nineteenth century therefore reasoned that the atom must be the smallest unit of matter.*It wasn't until the early years of the twentieth century that this concept began to break down, figuratively and literally.

*The word "atom" itself derives from the Greek word *atomos,* meaning indivisible.

THE "ORBITING" ELECTRON

Instead of a smooth, hard ball, the atom was discovered to resemble (metaphorically, at least) a tiny solar system with the sun at the center and planets revolving around it. In an atom, the "sun" is called the nucleus and the "planets" are known as electrons, though each orbit does not necessarily contain just one planetary electron.

The electron is one of several minute or subatomic particles that make up the atom. The most basic, stable particles that exist in nature, electrons are also the smallest. It would take one thousand million, million, million, million electrons to balance a one-gram weight.

So infinitesimal are electrons that they are really nothing more than bits of concentrated energy with a negative electric charge. Their mass, in fact, is the mass of the energy generated by their charge. That energy possesses weight, or mass, is a cardinal principle behind the atomic bomb; two of the factors in Einstein's now-famous equation, $E = mc^2$, are energy (E) and mass (m). Energy and mass coexist, slipping easily from one form to the other.

To explain an electric charge is extremely difficult—perhaps impossible. Suffice it to say that an electric charge is a basic unit of cosmic creation. If this sounds faintly mystical or religious, maybe it is. But while the concept of an electric charge must be accepted on faith, the behavior of electrically charged matter is well understood and easily documented. Electric charges, of course, occur in two forms, which for reasons of arbitrary reference are designated negative and positive.

Many readers will remember a high school general science experiment in which an amber rod was rubbed with fur or cloth and became capable of attracting strands of hair or small bits of cloth. (The word "electricity" derives from the Latin word for amber, *electrum.*) This exercise demonstrated the effects of the migration of charged particles—the friction rubbed some electrons loose and transferred them from one material to the other.

In his experiments during the 1700s, Benjamin Franklin observed that a certain kind of electricity seemed to move from object to object. He also noted that some substances stockpiled electricity in greater quantities than was normal, while others appeared to accumulate it in less than normal amounts. Franklin was the first scientist to refer to these types of objects as being positively or negatively charged.

Other substances (for example, rods of glass or sealing wax) can also be charged and given a force of attraction by being rubbed. Two such electrified glass rods suspended next to each other swing in opposite directions. Similarly, two electrified sealing-wax rods move away from, or repel, each other. But place a glass rod near a sealing-wax rod, and they attract each other. This phenomenon occurs because the glass rods are charged differently from the sealing-wax rods. Opposite charges attract; like charges repel. This basic physical law is the key that leads to a better understanding of the atom.

Let's go back to our solar system analogy in which we visualize the electrons as planets and the nucleus as the sun. Just as the mass of the sun dwarfs the planets, the nucleus mass is immense compared to its attendant electrons. And most of the mass of the atom (all but 0.05 percent or less) is that of the nucleus.

Yet by comparison to the total diameter of atoms, nuclei are very small. Most of an atom, in fact, is composed of the empty space that separates the nucleus from its "orbiting" electrons. Subatomic particles are found in the nucleus, and nuclei of the most complex atoms may contain as many as 250 of them. But all these particles lumped together yield a nucleus that is still 1/7,000 the size of a single atom.

Looking at these figures another way, if an atom were the size of a bowling ball, its nucleus would be only 1/400 of an inch in diameter —too small to be seen without magnification.

THE ATOMIC NUCLEUS

If electrons are negatively charged particles, and opposite charges attract, it follows that the nucleus of a whole atom should contain *positively* charged particles. This indeed turns out to be the case. The particles within the nucleus are called protons, and they do contain a positive charge.

The reason why virtually all the mass of the atom is packed into the nucleus is that protons are much, much heavier than electrons —between eighteen hundred and nineteen hundred times heavier. Despite this great disparity in mass, the positive electric charge of the proton is exactly equal in magnitude to the negative charge of the electron.

When a single proton and a single electron attract each other, the combination produces a specific atom—the hydrogen atom. The positive proton in the nucleus is exactly counterbalanced by the tiny electron circling it. As a result, the hydrogen atom is electrically neutral.

But more than one form of hydrogen atom exists. The second form, deuterium (also known as heavy hydrogen), is relatively rare compared to ordinary hydrogen. What makes deuterium heavy is the weight of something additional in its nucleus. This additional mass is accounted for by another type of particle called a neutron.

In mass, the neutron is roughly equivalent to the proton. It differs from protons and electrons, however, in that it has no electric charge. The name neutron derives from this electrically neutral state. Neutrons are contained in all the elements that exist in nature. Each element is a mixture of atoms with varying numbers of neutrons in their nuclei.

When different types of atoms of the same element exist, such as hydrogen and deuterium, the different forms are called isotopes. As a rule, all the isotopes of a particular element are equal in electric charge, because they all contain the same number of protons and electrons. Only the weight of the nucleus varies from isotope to isotope, because each contains a different number of neutrons.

With the basic rules established, let's see how they apply to a few simple atoms. The simplest, of course, is hydrogen (atomic number 1), because it contains one proton electrically balanced by one electron. Helium (atomic number 2) contains a nucleus of two protons and two neutrons with two electrons circling it in the same ring, or shell. Hence the atomic number of the various elements found in nature refers to the number of protons found in the nucleus.

A peculiarity of atoms is that only two electrons can be held in the first orbital shell. By the time we get to atomic number 3, lithium (three neutrons and the same number of protons in the nucleus), the three circling electrons require two orbital shells to accommodate them. The next orbital shell after the first can hold from one to eight electrons, while shells three to six can hold many more.

The subject of physics would be very simple indeed if all atoms were so natural-law abiding, such models of order and decorum.

Unfortunately for the academic averages of college students, this is not the case. Physical laws are very hard to pin down. As a matter of fact, they are still being discovered.

At one time scientists, struck by the resemblance of an atom to our solar system, assumed that it was held together by the force of gravity. Subsequently, they attributed the unity of the atom to the second fundamental force of nature, electromagnetism. But the Nobel Prize for Physics in 1979 was awarded to three men for work directed toward unifying these forces with two others, known as the strong interaction and the weak interaction. Both were discovered in the realm of subatomic activity. The strong interaction is thought to be what holds atomic nuclei together, whereas the weak interaction is responsible for the breakdown of certain atoms (as in radioactive decay).

If this was being written a few decades ago, Chapter 1 would have been completed many paragraphs back. The recent past, however, has seen research into the nature of matter pile one new discovery upon another.

Today we know that to discuss electrons, protons, and neutrons is merely to scratch the surface of atomic composition. Many other types of subatomic particles are now known to exist—roughly two hundred at last count.

Scientists are very precise people. Where others would have settled for the electron-proton-neutron atom and let it go at that, scientists retained little nagging doubts. If, for example, positively charged protons (which should repel each other) are packed tightly together in a nucleus, why doesn't the considerable strength of repulsion cause the nucleus to rip apart? Speculation grew that some special force kept protons together and operated only at extremely close range. Out of these notions grew speculation that a subatomic particle passes back and forth between protons and neutrons in the nucleus. Indeed it does. It was called a meson, from the Greek word for "intermediate," because its mass was thought to be somewhere between that of the proton and that of the electron.

But again the scientists doubted. Mesons answered certain questions, but they were too light to meet the requirements of their supposed function. As research continued, a heavier meson was discovered. To distinguish between the two, the Greek letters *pi* and *mu* were used. The heavier meson became a pi-meson, and the lighter one a mu-meson. (In scientific shorthand, they are usually referred to as pions and muons.)

But the fun was just beginning. Testing the hypothesis that a mirror image of the atomic nucleus exists, investigators found that every subatomic particle is matched by its opposite antiparticle—for example, an electron with a positive charge (the antielectron or positron) and a proton with a negative charge (the antiproton).

By the late 1940s the dozen or so known particles could provide answers to most of the remaining questions about matter, but not all. To help them keep the particles—known and suspected—in some kind of order, scientists divided subatomic particles by size into three categories: light (leptons), medium (mesons) and heavy (baryons). And on it went.

In 1956, a particle with hardly any mass and no charge was finally identified. This was the neutrino, and with it went all the neutrino variations. Light sometimes behaves as though it were made of particles; enter the photon. Certain particles that behaved atypically were called V-particles and were grouped together as hyperons. *Lambda*, *sigma* and *xi* variations of the hyperon were discovered. Finally the heaviest particle of all, called the omega-minus particle, appeared. Today physicists speak of families of particles and quarks—a theoretical grouping of particles that only a few of the top scientific minds completely understand.

In our discussion of radioactivity, we will deal with just the basic threesome: electrons, protons and neutrons. Our concern is with the human organism. And that's the direction we will follow after we lay the basic foundation for understanding radiation.

2
Atomic
Restlessness

The picture of the atom that we have presented is a still life. Atoms do not exist in that form any more than an airplane exists as a stationary object in midair, even though a photograph of the airplane in flight might create that very impression.

The essence of an atom is its constant restlessness. There is never an end to activity within, by, or upon atoms. This dynamic state produces various types of phenomena.

CHEMICAL ACTIVITY

We have established that an atom is the smallest unit of an element that retains the identity of that element. And we have described electron shells. Recall that the first shell holds a maximum of two electrons, the second and third a maximum of eight, and so on. It's the nature of an atom to want to be "fulfilled"—that is, to carry the maximum number of electrons in its outermost shell. When this is the case, the atom is very stable and will not react readily with other atoms.

Helium, for example, with one shell containing two electrons, has its outermost shell filled and is stable. Accordingly, it is the preferred gas for inflating dirigibles, because it will not explode. (An explosion is nothing more than a chemical reaction that takes place in a very short time.) Obviously, the outer electron shell is the critical one for chemical activity and the chemical stability of an atom depends on how many electrons are contained there. Because hydrogen carries just one electron, it constantly seeks a second to make itself complete. That's why hydrogen is so often encountered in nature as a pair of atoms. The nuclei share each other's electrons, with each behaving as though it has two of its own.

Atoms of higher elements (those with larger numbers of electrons) try to do the same thing but do not carry it off so neatly. Thus atoms of oxygen (eight electrons with six in the outer shell) also hook up and travel as pairs, or oxygen molecules.

Atoms of *different* elements can also hook up into molecules. These unions are called compounds, and a molecule is the smallest possible quantity of a compound. For instance, an atom of oxygen (six electrons in its outermost shell) will join forces with two atoms of hydrogen, and the two electrons supplied by these two hydrogen atoms brings the members of the oxygen's outer ring to the maximum of eight. What results is a molecule of the compound known as water. Molecules may be that small, or they may or contain thousands of atoms. The important thing to remember is that a molecule is the smallest quantity of a compound that can exist and still retain the identity of that compound.

Another point to keep in mind is molecular motion. Molecules in the earth's atmosphere, for example, move at speeds approximating a mile a minute. Even in solid materials, molecules pulsate rapidly. And if the temperature is raised, they move even faster. In gases or liquids where such movement exists, molecules collide with each other. Such an event can have several consequences. Colliding molecules may simply bounce off each other like the cars in an amusement park auto-dodger ride and go their separate ways. One of the molecules in a collision may sustain some injury, losing one or two atoms. Or, like cars that brush and then wear each other's paint, colliding molecules may exchange atoms. Or they may fuse to form one new, much larger, molecule.

Molecules may emerge from these random encounters with more or less than their usual complement of electrons. Such molecules, no longer neutral, will be positively charged if they are short some electrons or negatively charged if more than the usual number of electrons are present. The greater the disparity between what is and what should be, the greater the charge. These charged particles are called ions, and they are noteworthy for their ability to conduct electricity.

Salt dissolved in water will break down into its component ions— a positively charged sodium ion and a negatively charged chloride ion. This and other salts which ionize in water are called electrolytes, because they make it possible for the water to conduct electricity.

(This is a good reason not to swim in the ocean during a thunder-storm.)

ELECTRICAL ACTIVITY

These observations bring us to the second major phenomenon of atomic activity, electricity, and its handmaiden, magnetism. The two are very closely related and always coexist in nature. Electricity is nothing more than the movement of electrons. This can be demonstrated by the simple electric cell or battery, in which chemical activity is converted into electricity.

Such cells consist of a zinc rod and a copper rod inserted into a solution of strong battery acid. The acid not only dissolves the zinc, forming zinc ions, but is itself reacted on, forming its own ions. Because zinc ions are positively charged, their entry into solution leaves the zinc rod with an overabundance of electrons. Now, if the zinc and copper rods are connected with a piece of metal wire, these electrons flow through it to the positively charged copper rod. It's as simple as that. Whereas the random motion of migrating electrons are responsible for chemical phenomena, a stream of electrons moving in the *same* direction produces an electric current.

When a stream of electrons all move in the same direction, they create not only a current of electricity but also a magnetic field. To understand a magnetic field, remember the amber rod we talked about in Chapter 1. If you actually do the experiment, you will note that the attraction between electrically charged opposites (the amber and the bits of cloth) does not depend on their making physical contact. The gap between them constitutes an electric field. In general, an electric field always exists between any two charged particles of matter.

Magnetic fields resemble electric fields, but the space constituting the magnetic field is the space between the opposite poles of a magnet. The movement of an electrically charged particle (an electric current) generates a magnetic field. And the converse is also true. When magnetic fields are charged, an electric field is created. This connection between electrically charged particles and magnetism is crucial to the entire theory of electromagnetic radiation.

RADIOACTIVITY

In examining the atom, we have concentrated so far only on migrations of orbital electrons as they are involved with chemical and electrical phenomena. But the restlessness of the atom also carries through to the nucleus. The nuclei of some atoms are unstable. These nuclei are in a constant state of degeneration—breaking down and spewing, or "radiating," subatomic particles into the space outside the atom at great speeds. Elements the atoms of which contain such nuclei are designated radioactive, and we will discussing them at greater length a little further on. But for now let's just concentrate on the particulate matter that emanates from them.

When subjected to the force of a magnet, these radiations reveal themselves to be of three varieties. The path of one kind is deflected slightly in a particular direction. The path of the second is twisted sharply in the opposite direction. And the path of the third is unaffected by the magnet. These radiations eventually came to be called alpha, beta and gamma rays.

It was soon discovered that the alpha and beta rays are not rays at all, but streams of subatomic particles. Early in the twentieth century, *beta particles* were identified as electrons moving very rapidly. But haven't we said that there are no electrons in the nucleus? Actually neutrons, which are found in the nucleus, are combinations of protons and electrons whose electric charges cancel each other out. When an electron is radiated from the nucleus of an atom, it comes from a neutron, which at that moment becomes positively charged —in short, a proton.

Several years later, the *alpha particle* was identified. Because a magnet bent its path in the opposite direction from the beta path, the alpha particle was assumed to be positively charged. And it is much heavier than the beta variety, as evidenced by its smaller magnitude of deflection. These particles were found to consist of two protons and two neutrons in one tight little package. The gas helium contains a total of two electrons. The particle consisting of two protons plus two neutrons, or alpha particle, is therefore nothing more than a rapidly moving helium nucleus. This proton-neutron entity is extremely stable, and when an unstable nucleus undergoes change, it often throws the whole combination out as one unit.

Gamma rays, the third emanation, are quite another kettle of fish.

In fact, they're not particles at all. Gamma rays are extremely important in terms of their effects on living tissue. To understand this type
of emission more fully, we must delve into the subject of *electromagnetic* radiation.

If a stream of electrons is subject to a constant and regular change
in direction, an electric *and* magnetic field is created. It is called an
electromagnetic field, and energy transmitted through it is referred
to as electromagnetic energy or radiation. This energy does not flow
in a constant stream. It pulsates and oscillates in regular waves. The
concept of waves is important, because it helps us differentiate electromagnetic radiation from particulate radiation. (The latter involves
transmission of the kinetic energy of such atomic and subatomic
particles as electrons, protons, neutrons, heavy ions and alpha particles.)

An analogy can be drawn between the wavelike pulsation of electromagnetic radiation and a hypothetical automatic baseball pitching
machine that can toss out balls of different varieties at the speed of
light—about 186,000 miles per second. As these balls are propelled
from the machine, they bounce at regular intervals. Within a specific
time period, some bounce more often than others. You could think
of the distance between bounces as the *wavelength* (a characteristic
of electromagnetic radiation) and the number of bounces per second
as the *frequency* (another way to designate such radiations).

Because the distance between bounces multiplied by the number
of bounces per second always equals the speed of light, the higher the
frequency of an electromagnetic radiation, the shorter its wavelength,
and vice versa. The properties of electromagnetic radiations depend
on their frequency and wavelength.

Now that we have a way to distinguish between the various kinds
of electromagnetic radiation, let's see what they actually are. To do
this, we have to deal with the concept of *quanta*. Quanta are simply
units of energy—more like extremely small chunks—that vary strictly
in size with the type of energy. It's like cold cuts at a delicatessen,
available only by specific weights—one-ounce chunks of liverwurst,
two-ounce chunks of salami, three-ounce chunks of bologna, and so
on.

The key thing to keep in mind about electromagnetic radiation is
that it is not made of particles. Actually, it is easier to think of these
emissions as forms of light, and they all move at the speed of light.

Light, of course, takes up no space and is weightless. That's because it is not matter but a form of energy. The terms "electromagnetic radiation" and "electromagnetic energy" can therefore be used interchangeably.

One of the peculiarities of energy, or light, is that it behaves *as though* it were composed of particles the size of which is a measure of the energy they contain. You may remember from the first chapter that units of such light quanta are called *photons* (from the Greek word for light).

ELECTROMAGNETIC RADIATIONS

Electromagnetic radiations are of several types. Starting with the smallest quanta, we first encounter the waves used to send radio messages. *Radio-frequency radiations* are produced by the oscillation of a current of electrons in specific types of circuits. Within the family of radio-frequency radiations are several subclasses: in order of decreasing wavelength, they are the long, medium, short, and ultra-short waves.

Moving up the scale toward larger quanta, decreasing wavelength and increasing frequency, we come next to *microwave frequencies*. These particularly short radar waves lie somewhere on the scale between radio-frequency and infrared rays. Microwaves can be subclassified in order of decreasing wave length as ultra-high-frequency waves (UHF), super-high-frequency waves (SHF), and extra-high-frequency waves (EHF). Microwaves cannot usually be detected by the human senses. We can't feel any sensation of heat standing next to a microwave oven, but it cooks food just the same, and a lot more quickly than an ordinary electric oven.

Now we come to *infrared radiation* or radiant heat—radiant because heat can be felt without actually touching the source, such as a household radiator. Infrared light exists just below ordinary red light on the quantum scale (the Latin word *infra* means below). Light in this range cannot be perceived by the human eye, but when it strikes human tissue, such light is converted into heat that can be sensed. Of course, infrared heat lamps do give off a deep red glow. But this is ordinary red light; infrared light that creates heat is invisible.

Now comes another of those basic rules of physics. This one deals

ELECTROMAGNETIC RADIATION SPECTRUM (BOUNDARIES ARBITRARY)

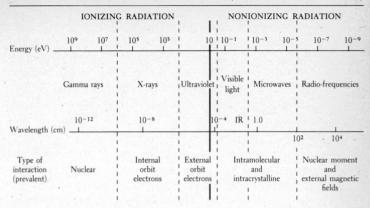

Adapted from S. Baranski and P. Czerski, *Biological Effects of Microwaves*, (Stroudsburg, Pa.: Dowden, Hutchinson, and Ross, 1976).

with the relationship between heat and atomic excitability. When subatomic particles are caused to move faster within a molecule (say, by friction), heat is generated. Conversely, heating a substance makes its atoms and molecules move more energetically. You can actually see this happening as water comes to a boil.

Remember that electromagnetic energy results from the vibrations of atoms and molecules, with the release of energy in the form of photons. The faster the vibrations, the more energetic the photons that are emitted. Continuing to raise the temperature of a body beyond the infrared wavelengths, at which it first gives off radiant heat, produces the first electromagnetic radiation visible to the naked eye—ordinary red light. That's why heating metal cause it to radiate heat first and then to glow red.

Red light contains the least energetic light quanta of all the visible light rays. It is therefore least likely to produce chemical changes. Whereas ordinary light alters the chemical composition of photographic film, the quanta of red light are insufficient to do this. Hence the small red bulb in a photographic darkroom produces light to see by but does not spoil the film.

If our red-glowing metal is heated further, its color changes from red to orange to yellow, as the frequency of the radiation and quanta increase. Additional increments of energy further increase frequency and progressively shorten wavelengths to produce green, blue, and violet light.

Let's stop and review for a minute. We've seen that the frequencies of such electromagnetic radiations as radio waves, microwaves, and infrared rays are too low to be visible. Ordinary light, with more energetic photons and higher frequencies, is visible as a series of colors. But atoms also send out radiations with too great a frequency to be seen by the naked eye. The first of these higher-frequency, shorter-wavelength, and more energetic emissions is ultraviolet radiation.

Just as the frequency of infrared light is a notch or two below that of visible red light, the frequency of ultraviolet light is slightly beyond that of ordinary violet light (the Latin word *ultra* means beyond).

At this frequency, photons of ultraviolet light are invisible, because the retina of the human eye is insensitive to them. However, these rays are sufficiently energetic to damage the eye; exposure to ultraviolet rays therefore necessitates wearing special goggles. Photons of ultraviolet light are also energetic enough to damage skin, which reddens and blisters after prolonged exposure. The familiar effects of sunburn are the result of ultraviolet light in the sun's rays. Another interesting aspect of ultraviolet light is its ability to affect photographic film, a property it shares with still more energetic forms of electromagnetic radiations.

The next more energetic radiation is the x-ray, a form of electromagnetic radiation with a greater energy and a shorter wavelength than that of ultraviolet waves. Electromagnetic energy often behaves like particles, particularly x-ray photons, which are so energetic that they can displace electrons.

A cue ball on a pool table is quite unlike the colored balls, but it can still propel them out of its path. Now, if a cue ball is able to do this to billiard balls, imagine what it could to to lighter and less dense ping-pong balls. This is somewhat analogous to x-ray photons and low-density, low-mass atoms. Human tissue, for example, is made up largely of atoms of hydrogen (one electron, and therefore very little mass), carbon (twelve electrons), nitrogen (fourteen electrons), and oxygen (sixteen electrons)—very few low-electron-density ping-pong balls on the pool table to impede the progress of the cue ball. Remember that atoms are mostly empty space. X-rays thus pass easily through the low-mass atoms of tissue.

But bones and teeth are a different story. They are composed primarily of phosphorus (thirty-one electrons) and calcium (forty electrons). X-rays do not penetrate them readily. This is the basis of the diagnostic x-ray picture familiar to anyone who has been to a dentist or suffered a broken bone. When a stream of x-ray photons is aimed at an x-ray film and a section of the body is interposed between the x-ray "gun" and the film, rays pass through the tissue with varying ease. When the film is developed, black areas indicate where x-rays passed through easily, gray where they encountered some resistance, and white where they were essentially stopped by more dense body parts. This allows bones beneath the surface of the body to be revealed clearly.

When heavy atoms, such as those in lead (eighty-two electrons), are placed in the path of x-ray photons, the pool table becomes clotted with colored balls, and the cue ball cannot jostle enough of them out of the way to get through. The density of lead makes it an ideal shield against x-rays.

Finally we return to gamma radiation, where our discussion of electromagnetic radiation began. The photons of gamma rays are very energetic, their wavelengths particularly short, and their frequency very high. Gamma rays differ from x-rays only in the manner of their production. Gamma rays are emitted by the spontaneous decay of radioactive substances; x-rays are produced by a machine.

Low-energy radiations (such as radio waves, microwaves, and infrared light) are produced by specific kinds of rotations of atoms and molecules. More energetic forms result from the movement of electrons within the atoms.

Remember the configuration of an atom, with electrons arranged in orbits, or shells, around a central nucleus? To understand how energy is given off or absorbed by these electrons, visualize them as being able to shift positions within their shells. When an electron moves from an outer to an inner orbital shell, it gives off energy as a photon. On the other hand, absorption of photons by an atom (assuming the correct size quantum) may cause an electron to leave an inner orbital shell and fly away from the nucleus. These movements are called energy shifts.

Now let's get back to our electromagnetic radiations. Generally, orbital shifts of outlying electrons produce lower-energy photons than shifts of inner-orbital electrons. Thus ordinary light rays result from energy shifts involving electrons situated farthest from the nucleus. Ultraviolet light is produced by energy shifts of electrons closer to the nucleus. And x-rays derive from energy shifts of the innermost electrons.

However, energy shifts may also occur in particles *inside* the nucleus. Remember our discussion of alpha and beta rays? When one of these particles is emitted by a nucleus, a rearrangement of the remaining particles within the nucleus occurs as the atom seeks to stabilize itself. During this process, tremendous photons of energy are given off, and these unusually energetic emissions are the gamma rays.

A few pages back, we mentioned that the paths of alpha particles and beta particles that emanate from atomic nuclei are bent in opposite directions when subjected to the force of a magnet. The third nuclear radiation—gamma radiation—isn't affected by a magnet at all. The reason, of course, is that gamma rays are not matter, but pure energy. Remember, however, that the more energetic quanta can act like particles and actually "shove" electrons around. This is what makes x-rays and gamma rays harmful to human tissue.

We'll pursue the effects of radiation on human tissue later on. But to put these effects into context, we must divide radiation into two more subtypes: ionizing and nonionizing radiation.

When a form of radiation interacts with an orbital electron, two results may be produced. If the photons possess amounts of energy no greater than the ultraviolet ray range, they simply produce an energy shift, raising the energy level of the orbital electron, without actually cutting it loose from nuclear control. This is known as an excitation. Though "excited" atoms or molecules are chemically reactive to an unusual degree (many chemical reactions take place only in the presence of ultraviolet light), they are not a major source of concern.

However, streams of high-energy photons (gamma and x-rays, for example) are sufficiently energetic to knock the outer electrons loose from their orbits, producing ions with a positive charge. Because of this phenomenon, gamma rays and x-rays are referred to as ionizing radiations. The rapidly moving, electrically charged alpha and beta particles can also bump electrons out of orbit when a collision occurs, and they too are considered ionizing radiations.

And what happens to the loose electrons and ions produced in the irradiated material? As you might expect, the electrons that are liberated stimulate further ionizations, usually thousands and thousands more than the original radiation caused. This propensity has great significance. Because ionizing radiation produces such a high degree of chemical reactivity, it can be extremely dangerous to living tissue, as we shall see in a later chapter.

Types of Ionizing Radiation

TYPE	CHARGE	DESCRIPTION	PRODUCED BY
Alpha	+2	Doubly ionized helium atom	Radioactive decay, primarily of heavy atoms
Beta	−1	Negative electron	Radioactive decay
Beta	+1	Positive electron	Radioactive decay
Protons	+1	Hydrogen nuclei	Generators and cyclotrons
Negative mesons	−1	Negatively charged particle with a mass 273 times that of an electron	Accelerators
Heavy nuclei	Have a range of charges	Any atom stripped of one or more electrons, and accelerated, is an ionizing particle. Deuterons and carbon atoms are examples.	Accelerators
Neutrons	0	Neutron	Atomic reactors, cyclotrons
Gamma rays	0	Electromagnetic radiation	Radioactive decay
X-rays	0	Electromagnetic radiation	X-ray machines and the rearrangement of orbital electrons

Adapted from Donald Pizzarello and Richard Witcofski, *Basic Radiation Biology,* 2nd ed. (Philadelphia: Lea & Febiger, 1975).

3
How Radioactivity Comes About and How It Is Measured

This chapter will apply some of the principles discussed earlier. We know, for example, that protons repel each other and would tear apart the nucleus of an atom were it not for the stabilizing influence of neutrons.

In atoms with few protons, an equal number of neutrons are enough to stabilize the nucleus. An atom with 4 electrons has 4 protons in its nucleus and 4 neutrons. When the nucleus contains 6 protons, these are matched by 6 neutrons, and so on. But not indefinitely.

When the number of protons is greater than 20, an equal complement of neutrons is no longer capable of keeping the nucleus together. Now the neutrons must outnumber the protons. And as protons accumulate and the atoms get heavier, the disparity between the number of protons and the number of neutrons increases. Elements with more than 30 protons need as many as 5 extra neutrons. For heavy atoms like lead and bismuth (with 82 and 83 protons, respectively), as many as 44 additional neutrons may be needed to stabilize their nuclei.

When an atomic nucleus contains more than 83 protons, its stability can no longer be maintained, no matter how many neutrons it contains. Now the nucleus is unstable and breaks down—not spectacularly, but simply by throwing off a subatomic particle every once in a while. Hence, in atoms with more than 83 protons, the stage is set for radioactive decay.

RADIOACTIVE DECAY

Uranium is an element contained in an ore called pitchblende, which occurs naturally in the earth's crust. The form found most often is the isotope containing 92 protons and 146 neutrons: uranium-238. The number after the element, 238, refers to the total number of protons and neutrons in the nucleus of that element. It is also called the atomic mass number. Because this isotope contains more than 83 protons, it is unstable. And because it is unstable, the U-238 nucleus exists in a state of constant breakdown, which is accomplished by the expulsion of an alpha particle.

This subatomic particle is basically a helium nucleus containing 2 protons and 2 neutrons. With the loss of 2 protons, that particular nucleus loses its uranium identity and becomes a nucleus of thorium, an element containing only 90 protons. The nucleus also contains 2 fewer neutrons, or 144. Thus the new entity is thorium-234.

This isotope is distinguished from the one usually found in nature by its extreme instability. The result is that thorium-234, too, breaks down. But in this case it does so by expelling a beta particle—that is, a fast-moving electron. Remember that, when an electron is radiated from the nucleus of an atom, it comes from a neutron. The loss of the electron makes this neutron positively charged and thus changes it into a proton. By gaining a proton, our thorium-234 nucleus becomes protactinium-234 (91 protons in the nucleus, but only 143 neutrons, for an atomic number of 234).

And it doesn't stop there. In seconds, protactinium-234 gives off a beta particle, which adds a proton and subtracts a neutron. Now our nucleus contains 92 protons, which is where we started, with uranium-238. But this time the nucleus contains only 142 neutrons. What we have is the isotope uranium-234, still unstable and still breaking down. This process is called radioactive decay: the spontaneous decrease in the number of radioactive atoms in radioactive material.

In the strange odyssey of uranium-238, we had traced it through thorium-234, protactinium-234, and uranium-234. Let's pick it up from there. You can probably figure out that, when U-234 emits an alpha particle, the result is thorium-230 (90 protons and 140 neutrons). Loss of another alpha particle reduces the totals to 88 protons and 138 neutrons, creating a nucleus of radium-226.

This breakdown process continues, involving radon (86 protons), polonium (84 protons), bismuth (83 protons), and lead (82 protons). When the number of protons finally drops to 82, the nucleus changes to that of lead-206, a very stable isotope that is not radioactive. And there the process ends. This entire evolution is known as a *radioactive*

Perhaps the most interesting aspect of radioactive decay is the production of energy. Every time a nucleus radiates a subatomic particle, it gives off energy. This is in accordance with certain basic laws of physics, which hold that matter and energy are simply different forms of the same thing and that neither can be created or destroyed.

Thus, when matter does change its form, some of it is converted into energy. On a simple basis, we can see this conversion in the combustion of coal or wood. Even though the ashes weigh less than the original substance, *nothing* has been lost. Some of the original material is contained in the ashes, while the rest of its atoms became components of gases that diffused into the atmosphere. When wood and coal burn, however, they also give off heat—a form of energy.

We know that energy is also produced when a radioactive nucleus changes form. But how much? According to Einstein, a fantastically large amount. When matter is converted into energy, the amount of energy produced equals the weight of the matter multiplied by a constant (the speed of light, or 186,000 miles per second) and multiplied by this constant again—the famous equation $E = mc^2$.

Thus, in the radioactive decay of uranium, the loss of weight is infinitesimal, whereas the production of energy is huge. And when a larger amount of uranium is made to break down suddenly, as it was at Hiroshima and Nagasaki during the summer of 1945, the energy produced can destroy a whole city. Even undergoing normal decay, radioactive substances can form highly energetic, powerful gamma rays and unleash wildly speeding alpha and beta particles. All three are capable of penetrating into and through human tissue like a dagger.

series. The steps in radioactive series can be traced through the isotopes of all radioactive elements, and each time the end result is common lead.

Radioactive decay is like a cascade of water pouring down over protruding rocks. It bounces repeatedly into the air, giving off a fine spray, and finally comes to rest as a quiet pool. Waterfalls obviously depend on a source of water, just as a radioactive series depends on a supply of radioactive material. But while a source of water can be replenished, the same is not ordinarily true of radioactive elements. Right now, the cosmos contains all the *naturally occurring* radioactive material there is or ever will be. When it is gone, it will be gone forever, and this statement could have been made when the universe was born untold eons ago.

HALF-LIFE

Will natural radioactive materials eventually disappear from the universe? The answer is yes . . . and no.

Radioactive elements decay at varying rates over periods of time ranging from billions of years to fractions of a second. Scientists refer to these time frames in a somewhat unusual way. Consider the radioactive element polonium (84 protons), a member of the uranium-238 series. If an ounce of polonium were to be sealed in a container for 100 years, only *half* of it would be left when the container was opened. The other half would have undergone decay to other material.

If we resealed the container and let it stand another 100 years, at the end of that time we would again find the amount of polonium reduced by half: a quarter of an ounce would remain. It would take another 100 years for the remaining quarter ounce to be reduced by half. What's more, no matter how small a quantity of polonium remained, it would still take 100 years for it to be reduced by half. Even if this process were to continue indefinitely, all the polonium could not be lost, because only half of what remained would decay to the next lowest element in the series during any given 100-year period.

Thus the absolute longevity of polonium cannot be fixed, though half its life (100 years) can be measured very accurately. For this reason, it is more accurate to speak of the longevity of a radioactive substance in terms of its half-life. *Half-life* is defined as the period of time required for half of any amount of an element to decay to the next lowest element in the series. Thus, regardless of the weight of the substance, physicists can always work with a constant number.

Just as it is impossible to determine when the last atom of polonium will disappear, scientists cannot tell the age of an already existing atom of this element. Some atoms might be in the process of creation as these words are written, while others may be hundreds of years old. No matter. The chances are fifty-fifty that any given atom will not disintegrate during the next 100 years. (These circumstances suggest instances of independent probability. It doesn't make any difference how many times you flip a playing card, for example. There is always a 50 percent chance it will land face-side-up on the next toss.)

Uranium (Radium) Series

ISOTOPE	HALF-LIFE	RADIATION	ENERGY* (MeV)
Uranium-238	4.5×10^9 y	α	4.18(77), 4.13(23)
Thorium-234	24.1 d	β	0.19(65), 0.10(35)
		γ	0.09(15), 0.06(7), 0.03(7)
Protactinium-234	1.18 min.	β	2.31(93), 1.45(6), 0.55(1)
		γ	1.01(2), 0.77(1), 0.04(3)
Uranium-234	2.50×10^5 y	α	4.77(72), 4.72(28)
		γ	0.05(28)
Thorium-230	8.0×10^4 y	α	4.68(76), 4.62(24)
Radium-226	1622 y	α	4.78(94), 4.59(6)
		γ	0.19(4)
Radon-222	3.82 d	α	5.48(100)
Polonium-218	3.05 min.	α	6.00(100)
Lead-214	26.8 min.	β	1.03(6), 0.66(40), 0.46(50), 0.40(4)
		γ	0.35(44), 0.29(24), 0.24(11), 0.05(2)
Bismuth-214	19.7 min.	β	3.18(15), 2.56(4), 1.79(8), 1.33(33), 1.03(22), 0.74(20)
		γ	2.43(2), 2.20(6), 2.12(1), 1.85(3), 1.76(19), 1.73(2), 1.51(3), 1.42(4), 1.38(7), 1.28(2), 1.24(7), 1.16(2), 1.12(20), 0.94(5), 0.81(2), 0.77(7), 0.61(45)
Polonium-214	160×10^{-6} s	α	7.68(100)
Lead-210	19.4 y	β	0.06(17), 0.02(83)
		γ	0.05(4)
Bismuth-210	5.0 d	β	1.16(100)
Polonium-210	138.4 d	α	5.30(100)
Lead-206	Stable		

*Numbers in parentheses indicate percent abundance.

Adapted from material in *Radiologic Quality of the Environment in the United States, 1977,* Office of Radiation Programs, U.S. Department of Commerce.

RADIOACTIVE ISOTOPES OF STABLE ELEMENTS

So far we have dealt only with naturally occurring radioactive elements that are always unstable, regardless of the number of neutrons in the nucleus. These are the elements with at least 83 protons in each atom. But what about elements with fewer than 83 protons? Are all isotopes of these elements stable? The answer is no.

One of the 12 most common elements in nature is potassium, which constitutes 1 percent of the human body. This element is basically composed of two stable isotopes. A little more than 9 out of 10 potassium atoms are potassium-39 (19 protons and 20 neutrons in the nucleus). Most of the remaining atoms are potassium-41 (19 protons and 22 neutrons). But there is a third isotope, potassium-40, and this isotope is so rare that it occurs only once in every 10,000 potassium atoms. As you can tell from its atomic number (the total number of protons and neutrons), each nucleus contains 19 protons and 21 neutrons. This form of potassium is radioactive and unstable; in fact, potassium-40 is the lightest radioactive isotope found on earth. (If the potassium in the human body contains some radioactive atoms, then each one of us is slightly radioactive already. Keep this in mind later when we discuss the effects of radiation on the human body.)

The half-life of potassium-40 is one billion three hundred million years, which means that most of it is already gone—converted to the stable element calcium-40 by the loss of a beta particle, which converts a neutron in the nucleus to a proton.

Other radioactive isotopes with atomic numbers of less than 83 can also be found on earth. Roughly 15 of them exist, including such common elements as carbon, phosphorus, and iodine. The half-lives of these isotopes are usually longer than that of potassium-40 and thousands of times longer than that of uranium-238. This indicates extremely weak radioactivity, and most of these isotopes decay to stable isotopes in one step.

PRODUCING RADIOACTIVITY ARTIFICIALLY

Knowing why radioactivity occurs naturally has enabled scientists to make it happen artificially. Formulating the theory was simple. Radioactivity involves the expulsion of subatomic particles from the nucleus of an atom, so all one had to do was arrange for these particles to be ejected. But to set up such a nuclear rearrangement, or reaction, was far easier said than done.

Before scientists could penetrate the nucleus and move things around, they had to get past the natural protective barrier of the nucleus and the atom's planetary electrons. These function like small, maneuverable destroyers or PT boats protecting an admiral's flagship. The small craft can intercept enemy vessels attacking the flagship. But they cannot stop bombardment by naval guns. If enough shells are fired, sooner or later one will score a direct hit on the flagship, causing pieces of it to shoot off in all directions.

Obviously projectiles were needed that could hit the flagship, or nucleus, with enough energy to cause the expulsion of matter. And what better source of such energetic particles than naturally occurring radioactive substances themselves. In other words, make a "gun" out of the subatomic particles released by a substance such as uranium. This could be accomplished by placing the naturally radioactive material in a container that was impervious to radiation but had one small opening. All the emanations would thus leave the container in a stream that could be "aimed" at a target atom to bring about a nuclear reaction.

The first successful nuclear reaction achieved by human manipulation was based on just this theory. Alpha particles (which are helium atoms containing 2 protons and 2 neutrons) were propelled into nitrogen (the nucleus of which contains 7 protons and 7 neutrons). The helium and nitrogen nuclei combined to form a new compound with a nucleus containing 9 protons and 9 neutrons. This conglomerate then emitted a proton. The result was two new elements, one with 8 protons in its nucleus (oxygen) and the other with 1 proton in its nucleus (hydrogen). In the course of this rearrangement, a certain amount of mass was converted into energy that was radiated in the form of gamma rays.

The reaction works fine for the lighter elements, despite the repulsion that exists between positively charged nuclei. Alpha particles move so quickly that they can overcome the repulsion of lighter elements with small positive charges. But when an element contains 20 or more protons in its nucleus, the positive charge is so great that even a fast-moving alpha particle is repelled.

Beta particles fare even worse. Because these electrons are negatively charged and very light, they are usually repelled by the fast-moving electrons that protect the atom. Gamma rays also fail to get the job done because they are without substance; they get through, but they are as effective as beams from the enemy raider's searchlight striking the admiral's flagship.

Finally, scientists decided to try protons. Large enough to elbow its way past protective electrons, a proton carries only a small positive charge (half that of an alpha particle). There is therefore only half the repulsion between protons and nuclei that there would be between alpha particles and nuclei.

Protons are easily obtained through the ionization of hydrogen. When the hydrogen atom is deprived of its planetary electron, what remains is the nucleus—a single proton. All that is required is a means of propelling this proton fast enough for it to slip through the electron blockade of another element and smash into the nucleus of that element with enough energy to jar a piece loose.

This can be achieved by bringing protons under the influence of high electrical forces. If the charged area is negative, protons move toward it, whereas a positively charged area sends protons careening away from it. The greater the charge, the greater the acceleration.

The first nuclear reaction involving accelerated protons was carried out on lithium-7 (3 protons and 4 neutrons in the nucleus). When an accelerated proton entered the nucleus, a compound nucleus containing 4 protons and 4 neutrons resulted. This was quickly reduced to 2 alpha particles (each with 2 protons and 2 neutrons). By now you will recognize these nuclei as those of helium-4.

After a nuclear reaction has been brought off, many of the products of the reaction do not immediately disappear but instead decay over periods ranging from minutes to years. Such radioactive substances are said to be artificial, but they differ from naturally occurring radioisotopes only in the length of their half-lives. Naturally occurring radioactive materials generally have much longer half-lives than artificial ones.

PARTICLE ACCELERATORS

Let's take a look at some of the equipment used to bring about nuclear reactions. Various means have been developed of hurling subatomic particles into nuclei with so much energy that the particles enter the nuclei and dislodge parts of them.

One of the first important particle accelerators was a machine called a *voltage multiplier*. This device multiplied electric potential volt by volt over a series of steps, each of which moved subatomic particles faster and faster. The energy of subatomic particles is reckoned in *electron-volts*. One electron-volt is the amount of energy gained by an electron accelerated by one volt of electric potential. Protons that were accelerated in the voltage multiplier sometimes possessed energy in the vicinity of 400,000 electron-volts. (The photons of ordinary light have energies of 3 electron-volts or less.)

A mechanism called the *electrostatic generator* is even more powerful. It works by building up positive or negative charges at the ends of the device, with increased electric potential in direct proportion to the magnitude of the accumulated charge. The electrostatic generator was the first of the so-called atom smashers to produce energies greater than a million electron-volts (abbreviated 1 Mev). And as improvement followed improvement, the device became capable of producing particles with energies as high as 18 Mev.

The principle behind the voltage multiplier is essentially the same as that behind the electrostatic generator. Acceleration is supplied in one burst, like the explosive charge in a bullet that sends it speeding on its lethal way. But acceleration can also be supplied in a series of pushes, each one adding more speed, in much the same way as a rolling hoop is made to move faster by someone running alongside and propelling it forward from time to time with a stick. By such means, particles can eventually be hurled forth fast enough and with enough energy to smash into atomic nuclei.

One machine that propels particles in just this manner is called a *linear accelerator* (sometimes referred to as a linac). Through a series of tubes of increasing length, particles are moved with ever-increasing speed until they reach the critical acceleration. Unfortunately, the tubes that finally kick particles up to the desired speed can be extremely lengthy—even miles long.

An athlete can run distances in a straight line or in laps around a circular track. So can subatomic particles. One way to save space and achieve the same result as the linac is to bend the tubes into circles. The device that operates on this principle is the *cyclotron*. Earlier models of this machine, less than a foot in diameter, were capable of producing particles with energies in the vicinity of 80,000 electron-volts.

These were followed by large units that could impart 10 Mev of energy and by modified and more sophisticated synchrocyclotrons with 800 Mev capabilities. Still later came the billion-electron-volt proton synchrotrons. These included the Bevatron (the unit called a Bev derives from the first letters in the phrase "billion electron-volts") and the Cosmotron, which suggests the power of the cosmos itself.

THE X-RAY MACHINE

Of course, the most familiar radiation-producing instrument is the x-ray machine. You'll recall from the last chapter that, the farther from the nucleus energy shifts occur, the lower the photon energies emitted. Ordinary light and ultraviolet radiation may be produced from shifts in outer-orbital or middle-orbital electrons, but x-radiation involves energy shifts of innermost electrons and produces photons of greater energy.

These innermost electrons are nudged into emitting x-rays when the atom is subject to bombardment by a stream of electrons the speed of which determines how deep they penetrate. The greater the speed, the deeper the penetration and the greater the energy of the x-rays emitted.

The x-ray machine you ordinarily see in your doctor's or dentist's office is designed to produce just such a stream of fast-moving electrons. These electrons strike a small dish of a heavy metal such as tungsten, disturb inner electrons of the tungsten atom, and cause them to be kicked out of their orbital shells in an energy shift that emits x-ray photons in the process. Some of the bombarding electrons also approach the *nucleus* of the tungsten, producing additional x-ray photons of varying energy. The x-ray beam produced in your doctor's or dentist's office therefore isn't homogeneous. It is a mixture of x-rays of varying wavelengths and degrees of penetrating capability.

TRACKING AND QUANTIFYING RADIATION

As we have seen, the more energetic x-rays can pass easily through dense matter like bone, which absorbs some of the x-rays and can be visualized on a photographic film. In the same way, the outline of the stomach and intestines can be viewed by filling them with a liquid that absorbs x-rays—the so-called barium meal eaten by patients about to undergo a gastrointestinal examination.

An x-ray film provides visual evidence that the body has absorbed radiation. Though human senses are unable to detect its presence, radiation produces certain physical and chemical effects that mark its passage in much the same way as the "invisible man" of movie fame did when he walked across snow or sand.

MEASURING DEVICES

Special instruments have been designed to reveal the "footprints" of radiation. Most radiation-detecting devices respond to the ionization produced. As photons interact with atoms, negative and positive ions follow the path of the radiation in numbers that vary with the amount of energy. These ions quickly recombine to restore a state of electric neutrality. However, if the beam of radiation is confined to a chamber with sides that are oppositely charged, negative ions migrate to the positively charged side, and positive ions to the negatively charged side. The charge built up on either wall can then be measured, and the number of ions formed in the wake of the radiation beam can be similarly determined. The latter measurement tells the amount of radiation that would be formed after the passage of a given beam of radiation through the air. An instrument of this type is known as a *radiation chamber* or *ionization chamber*.

Such chambers cannot themselves disclose particle numbers, amount of energy, path direction, or type of particle or photon, so the ionization chamber is usually fitted with some sort of counter. The most basic of these devices consist of a gas-filled chamber with a thin window area that a beam of radiation can easily penetrate. When a photon or particle enters the chamber, molecules of the gas are converted into ions, which of course makes them capable of conducting electricity. This electricity passes through a wire inside the container to a relay system that makes a clicking sound. Obviously, the greater the number of clicks, the greater the number of ionizing particles in the vicinity. This type of instrument was named after its inventor, Hans Geiger; it is the familiar *Geiger counter* used to detect radiation.

A counter that provides a more accurate estimate of particle radiation is the *scintillator*. As particles or photons pass through a certain crystalline material, they leave the usual ions in their wake. When these charged particles recombine, a tiny flash of light, or scintillation, results. The intensity of this flash is in direct proportion to the amount of energy deposited in the crystal. Should the crystal be of sufficient thickness to stop the particle, the intensity of the flash represents *all* the energy contained in the particle.

The passage of fast particles can be recorded by these counters in less than one ten-millionth of a second. And when fitted with electronic circuitry, scintillation counters can reveal the simultaneous passage of two particles within one-billionth of a second.

Whereas the standard free-air ionization chamber is the primary instrument used to measure radiation, the so-called *thimble chamber* is the secondary standard. This device offers a practical way to measure radiation, through calibration of readings from the primary free-air chamber. The thimble chamber is the instrument used most often to measure the output of x-ray machines.

To summarize, the various devices used to measure radiation are all based on physical and/or chemical effects produced by radiation. These include ionization, changes in temperature, production of light flashes, luminescence related to heat, and changes in the color of certain chemicals.

The most important of these effects are ionization of gases by radiation (ionization chambers and Geiger counters), production of light in various crystals (scintillation detection), and darkening of a photographic film exposed to radiation. This latter property forms the basis of x-rays used in medical and dental diagnoses and of the film badge—a light-proof film pocket worn as a badge by people who work in the nuclear industry or in medical, dental, or industrial radiology —used to measure exposure to ionizing radiation. The degree to which radiation causes the film to darken is an index of the dose the individual has absorbed.

UNITS OF MEASUREMENT

To assess the effects of x-rays on people, and for other purposes, scientists have devised units by which to measure the amount of x-rays individuals have been exposed to and the quantity of x-ray energy they have absorbed. The *roentgen* is the standard unit for measuring an x-ray exposure. Named after Wilhelm K. Roentgen, discoverer of x-rays, the roentgen is equal to that quantity of x-radiation necessary to form two billion ions in one cubic centimeter of air.

The roentgen is one of several important yardsticks of radiation. We will also have occasion to mention the *rad*. This acronym, made up of the first letters of the phrase "radiation absorbed dose," represents the amount of energy deposited in a gram of tissue by ionizing radiation such as x-rays. However, any particular exposure to radiation may produce varying numbers of rads in different tissues.

When other types of radiation are present, scientists use the term *rem* (from "radiation equivalent man"). This is an amount of radiation that would produce the same biological effects as a rad of x-rays. Where x-rays are involved, the rad and the rem are equal in value and can be used interchangeably.

The term *relative biological effectiveness* (RBE) is used to compare the amount of damage produced in tissues by various types of ionizing radiations. And in evaluating tissue damage, scientists speak of the *linear energy transfer* (LET). This is a measure of energy lost per micron of tissue along the track traveled by any ionizing particle. When radiation slams into living matter, its velocity is slowed and it loses energy as it penetrates. The deeper it goes, the more speed and energy it loses. The more slow-moving and highly charged the particle, the greater the frequency of tissue interaction and the greater its LET.

Think of it in terms of two metal balls, one heated to a glowing red color (charged), the second at room temperature (uncharged), both being propelled into a mass of marshmallow. The glowing metal ball is made to move much more slowly than the unheated one. As both balls pass through the marshmallow barrier, the heated, slower-moving ball does more damage to the marshmallow. This is analogous to the concept of LET. Highly charged and slow-moving radiations (for example, an alpha particle) cause more tissue damage than fast-moving, less highly charged radiations (such as electrons, protons, and neutrons).

We have talked at some length about theoretical principles, because they lay the groundwork for subsequent discussions. Now let's leave the realm of academic physics and see how this basic knowledge applies to the fragile human organism.

part two

Living with Radiation

4
Radiation: It's All Around Us!

One reason why we must be concerned about the amount of radiation introduced into our environment by human activities is that we are all exposed to varying degrees of natural radiation to begin with. Hence any sources that are "artificially" introduced add to radiation levels that are already present, perhaps in substantial amounts, in living cells. Before we examine the exposures to radiation that arise from our own activities, we should consider the sources and extent of this natural radiation background.

NATURAL RADIATION

COSMIC RAYS

On any given day, an individual simply following a normal routine is riddled by subatomic "grapeshot" penetrating every inch of his or her body by the thousands, every minute, without let-up. That this personal bombardment cannot be felt or sensed in no way reduces its effect.

The entire surface of the earth is subjected to the same never-ending hailstorm and has been since the birth of the cosmos. These penetrating projectiles are referred to collectively as cosmic rays, and they consist of atomic nuclei—mostly protons, but with a sprinkling of heavier particles. This is *primary radiation,* just as it arrives from space and before any interactions have taken place with atoms of the earth's atmosphere. After such interactions have occurred, more subatomic particles, including neutrons, are formed: so-called *secondary radiation.* Cosmic rays therefore consist of ionizing components and secondary, or neutron, components. Cosmic rays arrive on earth from three different sources. Galactic cosmic rays come from stars in the vast reaches of the universe—the major source of extraterrestrial radiation. These rays can travel such staggering distances that many now reaching the earth started in our direction before this planet even existed.

A second source is the radiation belts that surround our planet. These zones, named the Van Allen belts after their discoverer, follow the earth's magnetic field, in which are trapped large numbers of charged particles arriving from outer space. In closed patterns, the particles migrate in an endless stream, back and forth, from pole to pole.

The third source is the sun, which emits occasional bursts of radiation—the so-called solar particle events characterized by sunspot activity and solar flares, or geysers of flame, that shoot millions of miles into space. Flare activity is sporadic; peak periods were in 1948, 1959, and 1970. These cycles cannot be predicted. Flares usually subside within an hour, but high-energy particles and x-rays continue to be spewed out for many days. In addition to such bursts of particles, the sun emits a steady stream of ultraviolet light.

The enormous number of subatomic particles that reach the earth's surface represent only a minute fraction of the total number that enter the vicinity of the earth. The rest of this storm becomes trapped by the three layers of protection that swathe our planet. If it were not for the earth's atmosphere, ionosphere, and magnetic field, radiation bombardment would be so intense that life as we now know it could not exist. The ultraviolet light that emanates from the sun, if unimpeded, would alone be lethal to life on earth.

Starting at the surface of the earth, the first of these protective shields is the earth's *atmosphere*. Close to the earth's surface, the air is quite dense. Here subatomic particles and photons of energy arriving from space collide with molecules of the oxygen and nitrogen components of air, which prevent them from slamming into the surface. The atmosphere rises to an altitude of about 60 miles. Here air is so thin that its density is approximately one millionth of the surface density, and atoms of the gases that shroud the earth are so rare that they are virtually nonexistent.

The higher the altitude, the greater the cosmic ray intensity. For approximately every 10,000 feet of altitude, cosmic ray penetration increases by a factor of 3; thus it is 3 times greater at 10,000 feet than at sea level.

How does this affect professional airline personnel who spend long hours at altitudes in excess of 30,000 feet? Aircrew members absorb so much radiation that they are technically considered radiation workers. According to federal standards, radiation workers may receive an annual dose of up to 5 rems, and no more than 3 rems each calendar quarter. Most radiation workers receive only 30 percent of the allowable annual limit.

On a transatlantic flight from the United States to Europe, exposure to radiation may amount to about 5 millirems (about 1/200th of a rem). Aircrew members who regularly fly such routes are therefore limited by federal regulations to flying no more than 1,000 hours a year—a period of time during which they absorb about 1,000 millirems, or 1 rem, of radiation.

At altitudes of 50,000 to 60,000 feet, where the supersonic jets fly, the intensity of cosmic radiation is significantly greater than at the usual cruising altitudes of ordinary jet aircraft. But the actual time that supersonic jets spend in flight is considerably less than the flight times of subsonic craft, so radiation absorption is no greater. In the event of solar flares or sunspot activity, when atmospheric radiation levels were markedly increased, measuring devices carried aboard SSTs would alert the crews to drop down to a safer altitude and complete the flight at subsonic speeds. To date, this has never been necessary.

If altitude is a major factor in exposure to radiation, so is geographical location. Cosmic rays are deflected away from the equator, but they are channeled earthward at the north and south polar regions. Therefore, equatorial areas receive the fewest cosmic rays, and intensity is greater at the poles. When a charged particle or photon penetrating the upper atmosphere is intercepted by an air molecule, some of its energy is absorbed by the air molecule. Repetition of this process (called scattering) ultimately strips the charged particle of most of its energy, which is dissipated in the upper atmosphere and is seen as light—the auroras of the northern and southern skies.

Natives of tropical, sea-level cities receive minimal exposure to cosmic rays. But in areas of the United Kingdom, Canada, the northern United States, and Scandinavia, exposure to cosmic rays may be up to 1.5 times higher than in, say, Singapore. In high-altitude cities away from the equator, exposure is even greater. The intensity of cosmic rays in Denver, for example, may be as much as 120 millirems per year—3 times greater than it is in equatorial sea-level locales.

The accompanying table will give you some idea of how much cosmic radiation you may be absorbing in your own locale. Note that the table gives "whole-body" doses, not doses of internal radiation, which we will discuss later. Amounts shown are expressed in millirems per person, and they are estimated yearly averages. On page 68 of this book, you will find a chart designed to help you make a "personal inventory" of your own approximate yearly exposure to radiation. The table presented here is the first of several that you will use to fill out that chart.

The second of the earth's protective barriers—the *ionosphere*—is found at an altitude of between 35 and 200 miles above the earth. These layers derive from the breakdown and ionization of oxygen and nitrogen, and they consist of unorganized, free-floating electrons that intercept and reflect radiation of shorter wavelengths, like radio waves. This effect is what makes long-distance radio transmission possible on earth. Because it is nonselective, the ionosphere also reflects and deflects cosmic radio waves back into space. This is a distinct disadvantage to radio astronomers, who are keenly interested in studying the low-frequency radio waves emanating from stellar bodies.

Finally we come to the last shield that helps insulate the earth from cosmic radiation, the earth's magnetic field, or *magnetopause*. The magnetopause is not composed of substance as are the atmosphere and ionosphere. It resembles the force lines of a magnet. Imagine that the line between the earth's poles is a bar magnet, with attraction taking place only at the poles. The magnetopause is that area in which particles come close enough to the earth to be attracted—a force field reaching thousands of miles into space.

Keeping the bar magnet analogy in mind, you can see how cosmic rays approaching earth at the equator are repelled, whereas those coming in at the poles are attracted. Weakly charged particles that arrive from the sun in a steady, gentle stream (poetically referred to as the solar wind) are deflected by the earth's magnetic field. But more energetic cosmic rays knife through the field as though it were nonexistent. These rays are so energetic that they have been detected hundreds of feet below the earth's surface.

ESTIMATED EXTERNAL EXPOSURE TO
YOUR BODY FROM COSMIC RAYS

STATE OR TERRITORY	AVERAGE ANNUAL DOSE PER PERSON (MILLIREMS)	STATE OR TERRITORY	AVERAGE ANNUAL DOSE PER PERSON (MILLIREMS)
Alabama	40	New Jersey	40
Alaska	45	New Mexico	105
Arizona	60	New York	45
Arkansas	40	North Carolina	45
California	40	North Dakota	60
Colorado	120	Ohio	50
Connecticut	40	Oklahoma	50
Delaware	40	Oregon	50
Florida	35	Pennsylvania	45
Georgia	40	Rhode Island	40
Hawaii	30	South Carolina	40
Idaho	85	South Dakota	70
Illinois	45	Tennessee	45
Indiana	45	Texas	45
Iowa	50	Utah	115
Kansas	50	Vermont	50
Kentucky	45	Virginia	45
Louisiana	35	Washington	50
Maine	50	West Virginia	50
Maryland	40	Wisconsin	50
Massachusetts	40	Wyoming	130
Michigan	50	Canal Zone	30
Minnesota	55	Guam	35
Mississippi	40	Puerto Rico	30
Missouri	45	Samoa	30
Montana	90	Virgin Islands	30
Nebraska	75	District of Columbia	40
Nevada	85		
New Hampshire	45		

Adapted from material in *Radiological Quality of the Environment in the United States*, 1977, Office of Radiation Programs, U.S. Department of Commerce.

TERRESTRIAL SOURCES OF RADIATION

This brings us to our second source of radiation, the earth itself. When the earth was formed roughly 3–5 billion years ago, it coalesced from flaming radioactive gases—much the same material that the sun is composed of. As the gases cooled, the outer crust of the earth solidified, like the material that forms on the surface of soup as it cools in a kettle. The crust closed off the inside of the earth, preventing the cooling of molten solids and gases, which are trapped there to this day.

In the surface crust are several naturally occurring radioactive elements. Perhaps the most important of these primordial radionuclides are potassium-40, uranium-238, thorium-232 and (to a lesser extent) rubidinum-87. We saw in Chapter 3 that these radioactive elements decay and, in so doing, give off charged particles. This process accounts for the gamma radiations emitted by the terrestrial outer surfaces.

Certain regions, such as southwestern India, are known for high radioactivity. Beaches there are streaked with patches of radioactive sand composed of up to one-third thorium—the highest thorium content in the world. And natives of these areas are constantly exposed to radiation in doses greater than the standards set by the United States government for radiation workers.

The same kind of situation exists in areas of Brazil near the cosmopolitan city of Rio de Janeiro. In one small village of some 6,000 inhabitants, a joint study was carried out by the University of Brazil and New York University. Certain individuals were told to wear radiation-detection devices (camouflaged as religious medals, to ensure that they would be worn continuously). Records were carefully kept, and radiation dosage received in this village over extended time periods was noted. In these studies, investigators found that the exposed population received gamma ray doses in quantities 6 times greater than the world average and, in some cases, up to 30 times this quantity.

Other areas of the world with high levels of background radiation include France, Egypt, and certain tiny islands of the Pacific. This phenomenon is probably a result of regional geology. Certain rocks, for example, are more radioactive than others. Generally, igneous rocks emit more radiation than sedimentary rocks. But certain types of sedimentary rocks, such as shale, are unusually radioactive. Roughly three-quarters of the earth's surface is composed of sedimentary rock.

Levels of radioactivity thus depend on the type of rock as well as on the action of running water and soil build-up. Earth radioactivity is lowest in swampy areas with rich, black soil and highest in forested and arid lands.

The table on page 42 shows average amounts of whole-body radiation derived from natural terrestrial sources in various geographical areas. As a rule of thumb, doses in the Atlantic and Gulf Coastal Plains states vary from 15 to 35 millirems per year; in the northeastern, central, and far western regions, the average is 35–75 millirems per year; and on the Colorado plateau, average doses range between 75 and 140 millirems per year.

Summaries of estimated annual radiation doses received by individuals from cosmic and terrestrial sources represent the most recent information from the U.S. Department of Commerce. When you read the statistics, please remember that they are unremarkable. Levels of cosmic and terrestrial radiation have not changed significantly in the last 50,000 years. And these levels have probably remained constant for the last hundred million years.

Nor is geography a significant factor. The annual cosmic radiation dose in Colorado may be about 120 millirems per year, as opposed to only 30 per year in Florida, but the former does not even begin to approach the critical level. Neither do levels of terrestrial radiation, the thorium-rich sand of India and Brazil notwithstanding.

To date there is no clear-cut medical evidence to suggest that any one part of the world is more dangerous than any other, in terms of radiation. Even the highest levels of natural radiation in the world are well below those that begin to exert harmful effects on the human organism. Remember that we evolved to our current level of physical and physiological sophistication under unremitting exposure to cosmic and terrestrial radiation. Perhaps even because of it.

Radioactivity may, of course, also appear in materials that people have fabricated from natural sources. Thus those who live in cities may receive larger doses of radiation than their country cousins because of greater exposure to brick and concrete, which contain traces of radiation. The same is true of people who live in masonry or stone homes instead of houses built of wood. A leading expert on natural radiation has stated that the intensity of gamma rays from the radium in a brick house could conceivably produce up to three times the exposure anticipated in a wooden house.

INTERNAL RADIATION

With the soil of our planet giving off natural radiation, it follows that food and water that come either directly or indirectly from this

ESTIMATED EXTERNAL EXPOSURE TO YOUR BODY
FROM THE EARTH'S NATURAL RADIOACTIVITY

STATE OR TERRITORY	AVERAGE ANNUAL DOSE PER PERSON (MILLIREMS)	STATE OR TERRITORY	AVERAGE ANNUAL DOSE PER PERSON (MILLIREMS)
Alabama	70	New Jersey	60*
Alaska	60*	New Mexico	70
Arizona	60*	New York	65
Arkansas	75	North Carolina	75
California	50	North Dakota	60*
Colorado	105	Ohio	65
Connecticut	60*	Oklahoma	60*
Delaware	60*	Oregon	60*
Florida	60*	Pennsylvania	55
Georgia	60*	Rhode Island	65
Hawaii	60*	South Carolina	70
Idaho	60*	South Dakota	115
Illinois	65	Tennessee	70
Indiana	55	Texas	30
Iowa	60*	Utah	40
Kansas	60*	Vermont	45
Kentucky	60*	Virginia	55
Louisiana	40	Washington	60*
Maine	75	West Virginia	60*
Maryland	55	Wisconsin	55
Massachusetts	75	Wyoming	90
Michigan	60*	Canal Zone	60*
Minnesota	70	Guam	60*
Mississippi	65	Puerto Rico	60*
Missouri	60*	Samoa	60*
Montana	60*	Virgin Islands	60*
Nebraska	55	District of Columbia	55
Nevada	40		
New Hampshire	65		

*Roughly equal to the U.S. average.

Adapted from material in *Radiologic Quality of the Environment in the United States*, 1977, Office of Radiation Programs, U.S. Department of Commerce.

soil will also contain radioactive components. Foods grown in regions where radioactive levels are relatively high may contain correspondingly high levels of such radioactive materials as radium. Further, because oceans, seas, and lakes are exposed to cosmic and solar radiation, as well as to that emanating from the earth beneath them, food from aquatic environments may also contain radionuclides.

Eating and drinking, therefore, can expose us to *internal* radiation.

Ordinarily, radiation in food is not harmful per se. An extreme example is the Brazil nut, grown in Brazil, which is known for the high gamma ray activity of its soil. Yet this particular food source, which is one of the most radioactive—up to 14,000 times more radioactive than most common fruits, according to information made available by the United Nations—poses no threat to health by itself.

Fruits in general are known for their low levels of radioactivity, so they can be used as a standard against which to measure the radioactivity of some typical foods. Cereals, for example, can contain up to 600 times and tea 400 times the radioactivity of fruit. In the range of 100 to 150 times the level of fruit radioactivity, other "hot" foods include organ meats such as liver and kidney, flour, peanuts, and peanut butter.

By the way, these levels of radiation are only those contributed by alpha rays. Let's not forget about radioactive potassium-40, a minute component of regular potassium, but nevertheless a significant contributor to internal radiation exposure because potassium constitutes as much as 1 percent of the human body. Studies have shown that the radioactive potassium contained in human bodies contributes 10 to 20 times more internal exposure than any other ingested radionuclide.

Calcium, another element found abundantly in the body, can deviously raise levels of internal radiation exposure. For one thing, calcium bears a chemical resemblance to radium, and the radionuclide therefore shares calcium's ability to accumulate in bones. Calcium also allows itself to become contaminated with strontium-90, a radionuclide component of fallout from nuclear weapons testing. Because of its high calcium content, milk has been the subject of much publicity characterizing it as a strontium-90 carrier. Oddly enough, drinking less milk will not decrease one's exposure to strontium-90. The calcium in milk contains less strontium-90 than other dietary sources of calcium, such as vegetable matter. And if the body does not get its calcium supplies from milk, it will take them from these vegetable sources, thereby *increasing* the intake of strontium-90.

Radioactive materials occurring in soil naturally undergo the same metabolic incorporation into plant life. This matter eventually finds its way into the human body, either directly or indirectly, as a contaminant of the meat of cattle or other grazing animals.

The ponds, lakes, rivers, and oceans of the world support various species of minute marine life that are collectively called phytoplankton. Ecologically, the role of phytoplankton is to absorb nutrients in the aqueous environment and convert them into food for higher organisms, such as fish and other aquatic creatures. Unfortunately, the phytoplankton also takes up radioactive matter in the water. Marine life that feeds on plankton, therefore, becomes more radioactive. Most authorities agree that aquatic life—both salt- and fresh-water species—is generally resistant to radioactivity. Such life forms may function merely as carriers of radioactivity. And if we eat these forms of sea life, our levels of radioactivity increase.

Nuclear waste discharged into bodies of water finds its way into the human body by this route. Pollutants are absorbed by the plankton, which is taken up by marine life subsisting near the shore, where waste is usually dumped. When we eat shallow-water shellfish (such as clams, oysters, mussels, scallops, and crabs), radioactive elements may very well be deposited in many of our organs.

Some natural drinking water contains radon—a radionuclide resulting from the decay of uranium-238. In some instances, water used in homes may contain unusually high concentrations of radon. These concentrations seem to depend on geographical location. In rocky regions where the earth's crust consists primarily of granite, radon concentration in water may be high.

When radon is digested in drinking water, it is deposited in the stomach lining. But this radioactive element may also be diffused into the air in steam when radon-containing water is boiled. In this case, radon is deposited in the tissues of the upper respiratory tract. Scientists don't know for sure how widely water with high radon concentrations is distributed in the United States at this time. But it is clear that the degree of radon concentrations in drinking water should be investigated when high incidences of stomach and lung cancer exist in a particular geographical region.

TECHNOLOGICALLY ENHANCED NATURAL RADIATION

Radiation that derives from naturally occurring radioactive materials obviously cannot be avoided, but new and sophisticated technologies may enhance exposure by permitting greater concentrations of radioactivity to enter our environment. These sources are referred to collectively as technologically enhanced natural radiation. Natural gas brought up from wells, for example, contains high concentrations of radon and other naturally occurring radioactive elements. Fossil fuels such as coal and peat usually contain traces of uranium. And, in uranium mining itself, underground radioactive ores are transported to the surface, where they increase exposure not only to the work force involved in processing this material but also to the general public. A major factor in exposure to radioactivity is proximity to uranium mills or the by-products and waste of uranium ore processing.

Other types of ore may also cause problems. For example, uranium and thorium levels may be high when phosphates are being mined. This is particularly true in South Carolina. After mining has been discontinued, the land is often reclaimed and used for residential developments. These homes have been shown to exhibit levels of radon decay products that are appreciably higher than the levels found in homes not built on land reclaimed after phosphate mining.

Scientific advances may also increase exposure by permitting us to enter previously inviolate environments of concentrated radioactivity. Space travelers are extremely vulnerable to the radioactivity of the Van Allen belt and, outside the earth's protective layers, to cosmic rays. On more down-to-earth levels, people expose themselves to increased radiation by going into natural caves, underground caverns, and mines.

In the depths of uranium mines, radiation exposure results mainly from the presence of radon gas and the solid particles of its decay offspring, which for some reason are always designated as feminine: radon daughters. All too easily inhaled by miners, these particles may become embedded in the respiratory tract. And this delicate tissue can then be damaged through alpha radiation emitted by the trapped particles.

Thus our naturally occurring background of external radiation can also become internal, because virtually all of the significant radioactive nuclides somehow find their way into the human body. The most important of these are potassium-40, carbon-14, and (to a lesser extent) hydrogen-3, or tritium. These three elements are contained most plentifully in the human body, are involved in most vital biochemical processes, and are indeed essential to all forms of life.

Other radioactive elements that enter the body include thorium

No discussion of internal radiation is complete without mention of the self-imposed exposure of cigarette smokers. Among the naturally occurring radioactive elements present in cigarette smoke is polonium-210. Studies have shown that concentrations of this radionuclide are much higher in the lungs of cigarette smokers than in nonsmokers.

If polonium-210 is a daughter, its father is lead-210. This naturally occurring isotope is also found in cigarette smoke. In the lungs and bones of cigarette smokers, it appears in twice the concentrations seen in nonsmokers. These significantly higher concentrations are capable of inflicting as much as a 30 percent higher radiation dose on the bones of cigarette smokers than the dose received by the general population.

and uranium. These elements undergo decay in a series of steps that eventually result in the formation of lead. Many of these uranium and thorium daughters pass quickly through the body. But radium remains in the human skeleton, irradiating us perpetually from within.

With a few possible exceptions, internal radiation poses no more of a problem than external radiation. Just as we have been basking under an eternal shower of cosmic rays, we have probably been eating Brazil nuts and cereals while breathing radon gas since we started keeping time. Obviously occupational radiation workers, including medical technologists, nuclear power plant operators, and uranium miners, are at greater risk than the general population of developing certain illnesses. The exposures to which these people are subjected are closely monitored and regulated in the United States and abroad by government specification. Problems may develop, however, in particular locales where natural background radiation has been increased. We have mentioned seafood taken from waters containing radioactive wastes and housing built on the sites of mines or facilities in which radioactive material had been handled. These matters are presently under extensive scrutiny by the scientific community and the federal government.

Of particular concern (and the subject of much publicity) is disposition of the wastes, or tailings, from uranium mill operations. There is no doubt that people living at or near sites of former uranium-handling facilities receive high doses of radiation. At a tailings pile in Salt Lake City, Utah, for example, it has been estimated that exposed individuals receive a radiation dose to their upper respiratory tract of 14,000 millirems per year. Tailings piles in other locations also yield sobering statistics. In Grand Junction, Colorado, the figure is 8,100 millirems per year; in Mexican Hat, Utah, 1,200; and in Monument Valley, New Mexico, 900.

Surveys conducted by various environmental agencies clearly indicate that we should not reuse these sites without somehow eliminating the radioactivity and that, in any circumstances, tailings pile surfaces should never be used or developed. Such environmental surveys have been conducted in about 90 communities in the states of Arizona, Colorado, Idaho, New Mexico, Oregon, South Dakota, Texas, Utah, Washington, and Wyoming.

In studies designed to measure the radon given off by uranium mill tailings, it was determined that levels of the gas beyond half a mile from the tailings site were no higher than the normal levels for the community. However, a lot depends on wind direction; increases in radon levels were found to be higher at locations that are consistently downwind. Obviously, every effort should be made to measure radiation exposure to the general public living anywhere near a uranium mining and milling operation or, for that matter, near any phosphate and thorium mining operations.

Residents of areas known to exhibit high levels of technically enhanced natural radiation have been made aware of the condition by the local media and by various government, state, and municipal agencies empowered to deal with the problem. They are advised to avoid any additional exposure to radiation, and they are very carefully monitored.

For all other segments of the population, the yearly average radiation dose per person due to technically enhanced natural radiation is negligible. It need not be included in your personal radiation inventory.

HEALTH-RELATED HUMAN CONTRIBUTIONS TO RADIATION EXPOSURE

All told, external and internal radiation from natural sources brings the *average* human dose to approximately 100 millirems per year, or a lifetime dose of about 7,500 millirems. These doses alone would pose no problem. The potential for trouble develops only when we *add to* the natural radiation background. And this we do to an ever-increasing degree as scientific discoveries lead to technologic advances that greatly improve the quality of our lives, but also increase our exposure to ionizing and nonionizing radiation.

The second largest reservoir of radiation exposure in the United States, on both a total population and an individual basis, consists of health-related facilities—your doctor's or dentist's office, clinics, and hospitals where x-rays and radioactive materials are used in the diagnosis and treatment of disease.

Patients may be exposed to radiation externally through the use of an x-ray machine or internally by swallowing a radioactive substance, having it inserted into a body cavity, or being injected with it. Both external and internal methods of exposure are used to detect or control disease. X-ray machines usually subject patients to relatively low doses of radiation, but, when the doctor wants to visualize the inner parts of the body on a fluorescent screen (a process termed fluoroscopy), the same x-ray machine makes this possible by emitting much higher doses of radiation. Even higher doses are produced by radiation machines used in cancer therapy. Patients may also have sealed individual radiation sources implanted within the body—for example, radium implants for the treatment of cervical cancer.

Thus radiation may function as medication. And, like drugs, it can do good or cause damage, depending on the dosage. In 1974 about 317,000 patients with newly diagnosed cancer were treated with radiation. An additional 28,000 or so patients received radiation therapy for benign diseases.

Various forms of cancer are treated with extremely high radiation doses—6,000–7,000 rads (6–7 million millirads) in the case of bone cancer, for example. This radiation bombardment is restricted to small areas, in order to destroy just the tumor cells. Certain noncancerous diseases have been treated with radiation in doses ranging from 1,000 to 2,000 rads. In recent years, however, the use of radiotherapy for such benign diseases as bursitis, arthritis, ulcers, viral warts, and acne has declined in favor of newer, more effective methods of treatment. Here, too, extra care is taken to irradiate just the diseased area, because only a very narrow margin separates the dose needed for effective treatment and the dose above which normal, adjacent tissue is damaged.

X-RAYS

The primary contributor to medical exposure, however, is the diagnostic x-ray. A report released in 1979 by the U.S. Department of Health, Education, and Welfare contained some fascinating statistics. For example, approximately 145,000 dental x-ray machines are currently in use. In 1970 (the latest year for which demographic information on x-ray use is available), it was estimated that, of a population numbering roughly 200 million individuals, about 130 million, or 65 percent, were exposed to medical and dental x-rays. This figure represents an increase of 20 percent over 1964, while the population was increasing by only 7 percent. The use of x-rays clearly accelerated in the years between 1964 and 1970, and it has continued to climb in the intervening years. Today about 240 million x-ray examinations are done annually.

To a significant extent, this increase has probably resulted from blossoming technical capabilities, including the introduction of computerized axial tomography (CAT) scanners in 1973. Approximately 700 CAT scanners are now in operation and are being used on about 2 million patients a year. Current projections, based on the medical demand for CAT scanners, put the total number of these devices at 2,200 by 1982—about 1 for every 100,000 people. (We will discuss CAT scanners later.)

Apparently, the only factor operating to hold down the number of radiologic procedures performed in the United States is the shortage of trained personnel. Somewhere between 110,000 and 170,000 individuals work with x-ray equipment, but only about 80,000 are certified or licensed to practice.

Medical and dental x-rays probably account for 80 to 90 percent of all non-natural radiation exposure of the general public—somewhere in the range of 73 millirems per individual per year. The accompanying table gives estimates of the radiation doses absorbed by the bone marrow of our bodies in the course of typical x-ray examinations. This is an important parameter to look at, because there exists a strong correlation between the incidence of leukemia and the amount of radiation received by the active bone marrow—

that part of the body where the different types of blood cells are formed. The data presented in the table are grouped into high-, medium- and low-dose diagnostic and therapeutic procedures.

Keep in mind, however, that not all x-ray examinations are the same. A lot depends on how large an area of the body is irradiated. If the x-ray is confined to a small area, such as the liver, it will not have nearly the dangerous potential of the same radiation dose absorbed by the whole body over extended periods of time. The part of the body exposed is also significant in weighing risk. The mouth area can absorb over 900 rems with relatively little risk, while the same radiation level directed at organs of reproduction can prove quite harmful. These structures are more sensitive to radiation, as are the breast, the thyroid gland, and the eyes.

Obviously, measuring the effects of radiation depends on amount of exposure, size of the area x-rayed, exposure time, age of the patient exposed, and body structure. But it is even more complicated than that. According to one source, a radiologic examination of the gastrointestinal tract (the so-called G.I. series) doesn't stop there. It also produces radiation levels in excess of 1,300 millirems in the skin, 535 millirems in the bone marrow, and up to 90 millirems in the gonads, or reproductive organs.

When x-ray procedures like the G.I. series are medically indicated, the risk becomes subordinate to the benefits. However, a large proportion of the 240 million annual x-rays conducted in the United States every year may not be needed.

RADIOPHARMACEUTICALS

After diagnostic and therapeutic x-rays, the next largest category of radiation dosage through medical use consists of the radiopharmaceuticals, a class of compounds used in nuclear medicine. Radiopharmaceuticals are chemicals or drugs to which a radionuclide has been added. These substances (the so-called atomic cocktails) are then administered to patients as "tracers" to enable physicians to monitor chemistry and physiology within the body. These materials give off radiation that can be identified and measured by special devices. The energy of this radiation activates the instrument, and resulting read-outs enable technicians to distinguish between the tracer material and other, naturally occurring substances.

ESTIMATED EXPOSURE OF BONE MARROW TO
RADIATION DURING TYPICAL X-RAY EXAMINATIONS

TYPE OF EXAMINATION	MILLIREMS PER EXAMINATION
HIGH-DOSE GROUP	
Barium enema: lower gastrointestinal series	875
Pelvimetry: examination to evaluate the proportions of the birth canal	595
Barium meal: upper gastrointestinal series	535
Mammography: breast examination (exposure per breast)	500*
Lumbosacral spine: lower spine including spine tip	450
Small bowel series	422
Intravenous pyelogram: kidneys, ureter, and bladder	420
Lumbar spine: lower spine	347
Thoracic spine: middle spine	247
MEDIUM-DOSE GROUP	
Gallbladder	168
Abdomen	147
Ribs	143
Pelvis	93
Skull	78
Hip	72
LOW-DOSE GROUP	
Cervical spine: neck	52
Femur: upper leg	21
Dental: whole mouth	9

*Because of the widely varied techniques currently used in performing breast x-ray examinations, this number represents an approximate average of the mid-breast exposure produced by low-dose mammographic techniques.

Adapted from data in *The Selection of Patients for X-Ray Examinations*, U.S. Department of Health, Education, and Welfare Public Health Service, Food and Drug Administration, January 1980.

For example, blood usually contains sodium. If this substance is introduced into the bloodstream, it quickly mixes with the sodium already present and loses its identity. But if the injected sodium has been made radioactive, its atoms can be spotted and counted, regardless of what part of the body they have reached.

For this process to work, the tracer must act in much the same physiological and chemical manner as the substance it traces. In this way the body is "fooled" into processing the radioactively tagged material exactly as it handles the normal substance. Hence a molecule of blood containing radioactive sodium ions performs the same physiological function as an untagged molecule.

Radiopharmaceuticals can be used for both diagnosing and treating various medical disorders. In both applications, a key feature is the propensity of the isotope to home in and settle in a particular organ, often referred to as the *target organ*. Thus radioactive iodine beats a path to the thyroid gland, strontium heads for bone, and certain mercury compounds land in the kidney. In chemistry, this type of preference is termed an affinity.

One of the most common applications of this principle is in disorders of the thyroid gland, which during normal operations absorbs iodine from the bloodstream for subsequent manufacture of hormone material. If a thyroid disease is suspected, the patient can be given radioactive iodine, which, like untagged iodine, accumulates in the thyroid gland. Depending on how much of the tracer is detected in the organ after a specific time interval, a patient may be found to suffer from either an overactive or an underactive thyroid. The physical structure of the thyroid gland can also be assessed from results of the scan.

Unfortunately, some radiopharmaceuticals may have affinities for other than the target organs. These secondary organs are called *critical organs*. Even though they don't collect the major concentration of a specific radiopharmaceutical, the sensitivity of critical organs can play a role in determining what dose of a radiopharmaceutical can be administered. A certain dose may not be injurious to a target organ, but the proportion of the dose that goes to a critical organ may damage it. Remember, too, that these substances are also carried by the blood, inflicting a dose of radiation on the entire body as they travel through all body tissues.

How much exposure a patient can take depends first on certain physical variables (the type of radiation, amount of energy, and half-life of the radionuclide), which are known and well defined, and second on biological variables, which are extremely important in estimating radiation exposure but are less clearly delineated.

One biological variable that must be fitted into the equation is the physical condition of the patient. Most people who undergo nuclear medicine procedures are ill, and their target or critical organs may not be functioning well enough to handle what is considered a normal dose.

Then there is the age of the patient. Body functioning and organ size in children are different from those in adults. These factors may even vary between children and infants. Allowed dosage of radiopharmaceuticals obviously increases with increasing age. Standards exist for figuring exposure to adults, but no comparable standards exist at this time for infants and children. Risk to children is therefore harder to calculate, and there seems little doubt that nuclear medicine procedures may be more dangerous in children than in adults.*

So far in this discussion of radionuclides, we have covered only their use as an aid in diagnosis. To a much lesser degree, they are also employed in treating disease. This use represents only a small fraction of the total usage—perhaps as little as 3–5 percent. The most common of the diseases treated with radiopharmaceuticals are hyperthyroidism (overactive thyroid gland) and cancer of the thyroid.

Hyperthyroidism is treated and usually sucessfully controlled with iodine-131. However, this use has been associated with slight increases in the incidence of leukemia and thyroid cancer. The latter condition has also been seen to occur in children treated with radiation for enlarged thymus glands. At one time, iodine-131 was used only in adults over the age of 40, but the lower age limit is now 20 years. Pregnant women may not receive iodine-131 under any circumstances, because the radiopharmaceutical can pass into the body of the developing fetus and lodge in the thyroid of the unborn.

Doses of iodine-131 help slow the progress of thyroid cancer but at the same time deliver high radiation doses to the body's blood-manufacturing mechanism, the bone marrow. Thus treatment with iodine-131 may be limited by the development of leukemia, a disorder characterized by uncontrolled increase in one of the types of white blood cells. Here the potential risk of inducing leukemia by administering iodine-131 to treat a thyroid cancer must be weighed against the risk involved in not treating the thyroid cancer and allowing it to go unchecked. Many variables, including the patient's age and general state of health and whether the thyroid tumor has extended out of the thyroid gland into other areas, should be considered in the course of making this decision.

Use of radiopharmaceuticals is part of a trend that has seen the specialty of nuclear medicine become a vital and burgeoning field, along with the development of new radiopharmaceuticals, techniques

*At one time, radiopharmaceutical labels carried the warning that these substances should not be given to anyone under 18 years of age. But recently, more and more children have been given radiopharmaceuticals for purposes of diagnosis.

for using them, and sophisticated measuring instruments. Between 1960 and 1970, the use of radiopharmaceuticals increased fivefold. By 1980, such use was expected to be seven times greater than in 1970.

It has been estimated that, in 1980, the total whole-body exposure from radiopharmaceuticals to the entire population of the United States will be 3.3 million rems. That is a scary number, but remember that it represents a *total.* If you underwent no nuclear diagnostic procedure, your total was zero; if you did undergo a procedure, your average exposure was probably the equivalent of a dose rate of 3 millirems per year, according to a government report issued in 1979.

Before we leave the healing arts, consider the people who work in this field and whose exposure is therefore occupational. According to a report issued by the U.S. Department of Health, Education, and Welfare in 1979, about half a million individuals employed in medical and dental fields are exposed each year. Exposure occurs during such diagnostic or therapeutic procedures as giving x-ray examinations, administering radiation therapy, and preparing or injecting radiopharmaceuticals.

In addition, occupational exposure may occur during medical and dental research. Transportation and handling radionuclides can subject yet another group of individuals to radiation exposure. It is difficult for authorities to calculate dosage to medical workers, because the parts of their bodies exposed to radiation are not ordinarily considered.

According to government estimates, the average individual working in health care is probably exposed to radiation doses in the range of 60 to 100 millirems per year. Some estimates put the figures as high as 300 to 600 millirems. The latter figure already exceeds the yearly allowable dosage limit (500 millirems) for the public at large.

PRODUCTS AND OPERATIONS THAT EMIT ELECTROMAGNETIC RADIATION

Another source of radiation exposure is a kind of catch-all category that includes consumer products (sunlamps, TV sets and remote control devices, microwave ovens, smoke detectors, home and office security systems, luminous watches, and cardiac pacemakers); technical operations involved with products and services (high-tension electrical wires, mercury vapor lamps, lasers, and ceramic glazes); and military and commercial communications and radar operations (communications satellites, military radio communications, and radar detection systems). In the United States, the total average contribution to each individual's whole-body exposure from such sources is less than 5 millirems per year.

Many of these miscellaneous sources have been the subject of a great deal of publicity in the lay and scientific media. Citizens and public-interest organizations, for example, are concerned about the potential harmful effects of proximity to high-voltage transmission lines. Are rural dwellers who live near these crackling, humming wires strung on towers that march across the countryside in any danger? As of 1977, the biological effects of living close to a constant flow of energy in the 350-kilovolt range had not yet been determined.

Research is now under way in the United States to evaluate the potential biological effects of prolonged exposure to such radiation. This nonionizing radiation can damage the body, but its effects are directly mechanical rather than chemical. For example, ionizing radiation, which kills a cell via chemical changes leading to molecular rearrangement, is obviously more serious than the destruction of a cell via direct heat, but the differences are often academic.

LAMPS

Sunlamps are able to tan the skin because they generate ultraviolet rays. Nonionizing, to be sure, but these rays pack enough energy to destroy eye tissue. And this type of radiation is rendered even more dangerous by its invisibility.

It has been estimated that approximately 10,000 people a year receive sunlamp burns severe enough to require emergency hospital treatment. These injuries are usually traced to users who ignore or forget directions for the safe operation of these devices. Remember that using a sunlamp for one minute is equivalent to spending about an hour in the sun.

Another source of ultraviolet light is the mercury vapor lamp—that familiar form of illumination seen on city street corners and in sports arenas, gymnasiums, banks, and stores. Highly efficient and longer-lasting than ordinary electric lights, these lamps work by passing an electric current through a mercury and argon gas vapor, which glows as a result. Also produced during this process is ultraviolet radiation, which is prevented from escaping by a special boron-silicate glass shield. If this protective barrier breaks, however, the lamp spews forth unfiltered ultraviolet radiation as it continues to operate. Severe burns have been reported in people unknowingly exposed to broken mercury vapor lamps. To prevent such exposure to ultraviolet radiation, the Food and Drug Administration has set standards that require a self-extinguishing device that shuts the lamp off within 2 minutes after its outer glass shield is broken.

The huge tungsten-halogen lamps that provide illumination in television studios also produce ultraviolet radiation. Skin and eye injuries from ultraviolet radiation have been reported in people exposed to improperly operating light sources of this type.

MICROWAVE DEVICES

Radar detection is carried out in the microwave frequencies—the short wavelengths of electromagnetic radiation that fall just short of perception by the human senses. The shortness of these waves permits them to be reflected by larger objects, just as light is. But they can travel much farther, because their beams are not reflected or diffused by dust particles in the air. Special receivers pick up any of these waves that are reflected back by large, solid objects. An airplane, for example, can be detected within the area of radar wave transmission and its presence signaled by light release, or a "blip" on a viewing scope.

Microwaves are being used increasingly and may come to expose large segments of the population to nonionizing radiation. The entire country is blanketed by radar, as every airplane aloft is monitored by air traffic controllers. Radar also has applications in navigation, astronomy, meteorology, and law enforcement. And television remote control units operate on microwave frequencies.

In recent years, radar systems have also been adapted for catching shoplifters in retail stores. Merchandise is tagged with radar-detectable markers that are removed at the cashier's counter. Thus the shopper who patronizes stores equipped with these anti-theft devices receives that much more exposure to radar.

Microwave technology is proliferating to such a degree that it is rapidly becoming a significant source of radiation exposure in the

world. Modern communications both private and military, for example, rely heavily on microwave frequencies. Telephone and television
signal-relay towers dot the landscape like confetti on the floor at a
New Year's Eve party. Other relay stations connect call boxes for
motorists' use along highways or link up private computers that can
"talk" to each other from city to city. Certain burglar alarm systems
operate on radar frequencies, and so do automatic garage door openers. Then there are CB radio transmitters—about 15 million of them
—broadcasting from cars, trucks, and homes all across the country.

Above the earth, microwaves are broadcast from the troposphere
by powerful transmitters that serve the civilian and government satellite communications systems. Two extremely powerful microwave
broadcasting satellites, established in fixed orbits since 1979 and
using frequencies in the range of billions of cycles per second, penetrate the ionosphere to relay telephone conversations, TV and radio
broadcasts, and diplomatic communications to receivers around the
world.

The recourse to microwave technology by the U.S. armed forces
is, of course, a closely guarded secret. But it must be considerable in
light of extensive use for tracking stations, scanning radar, range
finders, guidance systems for the nuclear fleet, and other complex
and sophisticated equipment.

Perhaps the single most important application in recent years has
been the microwave oven. As the use of these convenient devices has
increased, so has concern for the safety of individuals who use them.
The radiation emitted by microwave ovens is nonionizing, but the
danger is still considerable and worthy of the attention that news
media and consumer advocates have paid it.

To understand the hazards inherent in microwave cooking, consider the reactions that can take place when microwave energy comes
in contact with matter. Depending on the frequency and wavelength
of the microwave and on the molecular structure of the material
being irradiated, three reactions may occur, individually or collectively. The first of these is *reflection,* in which energy impinging on
the surface of the substance is rerouted away from it. Second, *transmission* may take place; here, the radiated energy generated passes
unchanged through the irradiated substance. Third, *absorption* may
occur, so that energy is dissipated *within* the substance rather than
being reflected away from it or passing through.

You may recall from Chapter 2 that nonionizing radiation simply raises the energy level of an orbital electron, causing an excitation rather than a breaking loose of the outer electron from its orbit. When a material is acted on by a microwave oven, the molecules are excited by the waves of energy that penetrate the material. Excitation leads to rapid movement of the molecules (remember how water comes to a boil) and, of course, to the generation of heat.

Just how much heat is generated depends on such factors as the size of the microwave generating system, the structure of the material to be cooked, and its temperature as it goes into the oven. The type of heating to be done determines how much energy is required. Lower oven settings are used to reheat food already cooked; more power is required to cook raw or frozen food.

The size, shape, and density of the material to be cooked are also key factors in the amount of irradiation required. As in regular cooking, the thicker the cut of steak, the more time required to cook it.

Readers who have used a microwave oven are well aware of one of its major advantages: the speed with which food can be prepared. That's because microwave cooking is much more efficient, heating only the food you are preparing. In a conventional oven, heat must build up throughout the cooking chamber until an adequate temperature has been reached. At this point, the surface of a roast begins to be heated and then, by conduction, its center heats up. One problem is that heat must build up until the center begins to cook, and if too much heat is generated, the surface of the meat burns. While heat is building up, the inside of the oven can scorch, its exterior surfaces can get hot enough to burn an errant hand, and eventually the entire kitchen can become unpleasantly warm.

In a microwave oven, conversion to heat begins as soon as the microwave energy starts to penetrate the meat. This penetration occurs to a depth of only 2.5 to 3 inches below the surface, but the interior of a larger and thicker roast cooks via conduction.

Our roast receives its microwave energy directly from the source and by reflection from the sides and bottom of the oven. Heating of the roast is therefore more uniform. The interesting feature, however, is that the sides and bottom of the oven do not become hot. That's because they don't absorb the microwave energy, their molecules don't become excited, and heat is not generated. When using a microwave oven, you can cook in any vessel that permits the *transmis-*

sion of microwave energy—that is, lets it pass through unchanged. Containers can be made of glass, china, plastic, or paper; they don't absorb or reflect microwave energy. Metal containers are not suitable; they reflect the energy and keep it away from the food.

Obviously, what microwaves can do to a roast they can do to human flesh, which is also protein and similar in molecular structure. Though microwave ovens emit radiation of the nonionizing variety, damage is still possible. As a single source of radiation, these devices may not yet be significant, but their increasing use suggests a problem for the future. We will have more to say about how microwaves affect living tissue, and how the benefits of microwave cooking can be derived with minimal risk, in later chapters.

PRODUCTS THAT EMIT IONIZING RADIATION

Among products that emit ionizing radiation are watches and clocks. The faces of timepieces designed to glow in the dark are painted with paints containing minute amounts of radium. As the radium gives off radiation, its energy causes the phosphorescent paint to glow.

The radiation produced by radium and its decay products is energetic enough to pass easily through a watch crystal. For this reason, the use of radium in timepieces is declining. Tritium has largely taken its place. This radionuclide gives off only low-energy beta rays, which pass less easily through glass or plastic. Certain types of digital watches also contain tritium, and these watches are constructed so that breakage of the tritium container is extremely unlikely. Even if exposure to the entire tritium content of a watch were to take place, such a radiation dose would amount to less than half a millirem per year.

Modern smoke detectors also contain radioactive materials. In these devices, radiation ionizes the air in a sensing chamber, generating a flow of electricity. When smoke enters this chamber, it affects the flow of ions, which sets off an alarm. Originally, radium-226 was the preferred radionuclide for smoke alarms, but the most widely used substance today is americium-241. With this substance, exposure is virtually nil.

The use of uranium as a coloring agent dates back as far as the eighteenth century. In various chemical combinations and salts, it was used to glaze glass, enamel, and porcelain with a variety of colors and fluorescent hues. One of its major uses was in bathroom sets, making it possible to bathe oneself in radiation both figuratively and literally. In the early 1970s, the U.S. Food and Drug Administration acknowledged a spate of adverse publicity by prohibiting the use of uranium compounds as color additives. However, dishes and other tableware glazed with uranium pigments can still be found in antique shops as collectors' items.

The use of uranium compounds for coloring has also found wide application in dentistry. For better than half a century, the manufacturers of false teeth have incorporated uranium pigments into their products in an attempt to duplicate the natural coloration and fluorescence of real teeth under all lighting conditions. It is thus possible to chew on radioactive food with radioactive teeth.

While the risks involved in uranium-glazed porcelain teeth are not known, concentrations of the radionuclide are so low that the government chooses not to regulate this use. However, even though no significant health hazard is deemed to exist, the dental industry has been urged by the Bureau of Radiologic Health of the U.S. Food and Drug Administration to find nonradioactive substitutes within a reasonable time. Until suitable substitutes are found, a maximum permissible concentration of uranium has been set. It is expected to keep dosage within the limit (1.5 rem per year) set by an international commission on radiation protection.

LASERS

One of the most exciting discoveries of modern times can produce radiation exposure. In fact, "radiation" is part of its name: light amplification by stimulated emission of radiation—or simply *laser*. This remarkable device is based on the fact that, under highly specific conditions, atoms can be made to emit light waves that are completely coordinated. (Imagine a chorus line with all the dancers in step and kicking in unison.) This is not the case with all other light sources, in which atoms act independently and the light given off is a heterogeneous mixture of electromagnetic waves (each member of the chorus line dances without reference to any other dancer). Such light is unorganized and not particularly strong. But when all the atoms "kick in unison," each wave reinforces the others and an extremely intense and unified beam results.

This, basically, is the laser. It can vary in power by millions of watts and operate continuously. Or it can pulsate with millions of watts of power in each pulse, enough to perforate steel and rock. But the key feature of a laser is its coherence. Unlike the light from a flashlight, which diffuses as it travels farther and farther from the bulb, the laser beam maintains its dimensions at every point along its path, with no loss of energy at any point.

Laser light can be sent in a pencil-thin beam to distances never before possible (to the moon and back, for example), making distance measurements more precise than ever dreamed of. And laser light is brighter than all other known light; it is two orders of magnitude brighter than the sun!

The guts of a laser is a glass tube, which contains helium and neon gas, and glass plates at each end. These act as mirrors, except that one is coated lightly, so that a very small percentage of light striking it from inside the tube "leaks" through. When an electric charge of very high frequency is passed through this gas mixture, the helium atoms become excited and collide with neon atoms. In a complex series of actions and reactions, the neon atoms give off electromagnetic radiation that can be controlled and emitted in synchronized waves.

The process is steadily repeated and, as waves are reflected back and forth within the tube, they stimulate other neon atoms to emit photons of radioactivity, building up one very strong, unified wave. The small portion of radiation that leaves the tube through the slightly leaky mirror is the laser beam. The neon atoms have added all their radiation together so that the waves are all "kicking in unison." And, as with the famed Radio City Rockettes, the effect is spectacular.

Because lasers can have different wavelengths (but only one within each laser), the longer the wavelength, the greater the penetrating ability. Lasers can emit light of any wavelength, from infrared through the spectrum of visible light to the ultraviolet ranges.

The organ most subject to damage by laser is the eye. Ultraviolet radiation is absorbed by the cornea and lens. But visible light passes through the cornea and the lens, which magnifies its power and focuses on the retina a quantum of energy significantly greater than that which left the laser.

Laser damage can be serious. However, the various potential scientific uses for the device are unlimited, and the benefits to be derived from its use greatly outweigh the known risks. We will enumerate their liabilities and benefits later; but for now, suffice it to say that lasers do not yet represent a significant source of radiation exposure.

ENOUGH LITTLE THINGS MEAN A LOT

Let's leave the subject of light for now and go on to some objects encountered every day that are sources of measurable radiation. Your TV set, for example, is basically a low-voltage piece of x-ray equipment. From within the picture tube, electrons are emitted by a cathode (actually referred to as an electron gun) and shot onto the phosphorus coating of the tube that emits light. (If you bring your arm close to the surface of a TV picture tube while it is operating, you can actually feel the barrage of electrons.) This is basically the principle of the x-ray machine. But TV sets use much lower voltages. Furthermore, radiation from a TV set must pass through glass and plastic shielding. And unless you sit very close to the TV set, distance from the electron gun also serves to dull its sting. As a source of radiation, TV sets probably contribute less than 1 millirem per year to total radiation dosage for the average viewer.

Another fairly common device that can make a small contribution to radiation exposure is also based on the electron gun principle. This is the video display terminal hooked up to computers in banks and other institutions. With this device, your checking account balance or other information can be "punched up" and displayed on a screen for the operator to read.

Basically, the video display terminal, or VDT, operates almost exactly like a TV set. It contains a source of electrons (the cathode) and a phosphorus-coated screen (the anode) inside a specially designed vacuum tube. Under high voltage, the gun discharges a spray of electrons onto the tube in accordance with electronic signals. When these electrons interact with the phosphorus inside the screen, visible radiation is formed and appears as a display of letters and symbols.

Several types of electromagnetic radiation may be produced by VDT devices. If voltage is sufficiently high, for example, low-energy x-rays can be generated by the cathode ray tube. Then there is the phosphorus coating of the display screen, which can emit ultraviolet, visible, and infrared radiation. Certain of the electronic components and circuits can produce radiation in the radio frequency range.

Back in the area of medically related radiation sources is a device that can be life-saving, and the benefits of which far outweigh any risk. This is the cardiac pacemaker, used extensively to regulate heart rhythm through electrical impulses delivered directly to diseased muscles in the heart. These devices are put in place surgically, under the skin of patients with certain types of abnormal heart rhythm. Electrical impulses are carefully timed to match normal heart rhythm. They cause the muscle to contract normally, which it would not do if the pacemaker were not present.

Conventional pacemakers are powered by mercury batteries that usually last from two to three years. When this power source begins to taper off, the entire pacemaker must be changed, again surgically. In 1972, however, a new type of nuclear battery containing plutonium-238 became the subject of study. This investigation was a limited one. But the nuclear-powered pacemakers were implanted in hundreds of patients who are benefiting from the ten-year-plus life span of these batteries.

In 1976, the Nuclear Regulatory Commission's Office of Nuclear Material Safety and Safeguards concluded that the circumstances warranted use of the batteries and licensed plutonium-powered pacemakers for routine prescription.

Use of cardiac pacemakers is now widespread. Because the operation of pacemakers depends on the production of electricity from a radioactive source, other pieces of equipment that also involve the production of nonionizing radiation can cause what is known as *electromagnetic interference* and produce a wide range of malfunctions. For example, the microwave oven can be a real hazard to individuals wearing cardiac pacemakers. Any number of incidents have been recorded in which the radiation emitted by a microwave oven disrupted the usual rhythm of a pacemaker and nearly caused the death of the wearer. This is why people with cardiac pacemakers are usually warned to avoid areas in which microwave ovens are operating. In hospitals where microwave ovens are often used to heat food, the risks can be magnified considerably.

Nor are microwave ovens the only bane of cardiac pacemakers. Any number of other microwave sources—broadcasting stations, radar amateur radio antenna stations, police car or taxi UHF radios, and

Many operators of video display terminals have complained of such symptoms as eyestrain, loss of visual acuity, frequent headaches, and (occasionally) alterations in color perception, loss of sensation in fingertips, nausea, sensitivity to temperature and noise, and general fatigue. For this reason, a study was conducted by the National Institute for Occupational Safety and Health of the Department of Health, Education, and Welfare to determine whether ultraviolet light, visible light, and infrared radiation, as well as radio frequency radiation and x-rays, were being emitted by presently available video display terminals.

According to this government bureau, a previous examination revealed that x-ray levels emitted by VDT units rarely exceeded the level of ionizing radiation that is present naturally. It now appears that, although some x-radiation is given off, the levels are well within normal tolerances. On the basis of these standards, the government expressed doubt that the VDTs examined were capable of producing radiation levels high enough to constitute an occupational risk to the eyes.

Further, the amount of visible light, or fluorescence, produced by phosphors inside the screens of VDTs through excitation by ultraviolet radiation or x-rays could not be detected by available instruments. And finally, no measurements taken in any way suggested that older VDT units, perhaps functioning improperly, could produce increased radiation levels of any type. This report concluded that the eyestrain and other complaints reported by VDT workers were probably valid but might be caused by other factors in the employee's environment.

CB radios—can also throw off microwave radiation and interfere with cardiac pacemakers or other sensitive electronic equipment. Everyone is probably familiar with the restriction on two-way radios near building sites where dynamite is being exploded via an electric detonator. The joke about a pacemaker wearer who sneezed and opened his garage door is not so very far removed from fact.

As a source of radiation, the plutonium-238-powered pacemaker contributes a fair amount of exposure to the wearer—about 5,000 millirems per year. But the risks involved in repeated surgical implantation of mercury battery pacemakers every two or three years are considered potentially higher than the risk associated with this radioactivity.

Exposure isn't limited to the patient, however. Suppose that a man wears a pacemaker. His pacemaker also exposes others near him to radiation. His spouse, for instance, receives a total body dose each year of approximately 7.5 millirems. Other household members are exposed to slightly more than 1 millirem per year. And work associates receive a dose of just under 1 millirem per year. This may not sound like much, but when you consider that the average person probably associates with about 25 people during her or his daily activities, it adds up to a lot of radiation.

NUCLEAR TESTING

The last important source of radioactivity is the nuclear establishment: national governments engaging in nuclear tests and private industry involved in what is known as the uranium fuel cycle. This term refers to all operations involved, from the initial mining and milling of uranium ore, through use of the radioactive material in power generation, to the disposal of radioactive waste. This subject is important enough to merit detailed coverage, which we will provide in the next chapter.

For now, suffice it to say that commercial nuclear power is not ordinarily a significant source of radiation exposure. According to statistical information released by the federal government in 1979, the amount we pick up from the environment is equivalent to 1 millirem per person per year. Exposure for individuals who work in the nuclear power industry is equivalent to an additional 0.15 millirems per year. But remember that these figures are applicable only under normal circumstances. It is abnormal circumstances that seem to worry people. We'll have more to say about this as well.

As for nuclear testing, significant doses of radiation resulted from weapons tests conducted in the atmosphere by the United States from 1945 to 1962, and since that time by the U.S.S.R., the United Kingdom, France, and China. Those tests produced what is known as *fallout*—radioactive airborne particles that settle to the earth's surface after a nuclear explosion. In the area of a nuclear explosion, fallout reaches the ground within about 24 hours. But some of this pariculate matter drifts into the troposphere, a layer of the earth's atmosphere below the stratosphere. In the troposphere, the material is wafted across the face of the earth, roughly at the latitude at which the explosion took place, and drops to the ground along a narrow path. Radioactive matter also reaches the stratosphere and then falls out over extended periods of time, blanketing large portions of the earth's surface. Tests conducted in the atmosphere have produced fallout destined to shower the global population with low-level radiation at a decreasing rate for years to come.

The atmospheric testing program conducted by the United States Atomic Energy Commission took place at sites in Nevada and in the South Pacific Ocean. Predictably, localized fallout subjected to radiation exposure large numbers of people living in eastern Nevada, southwest Utah, northern Arizona, certain Pacific islands, and areas downwind of these sites.

During the 17 or so years of this program, approximately 250,000 military and civilian personnel and 150,000 people employed by the Atomic Energy Commission participated. Roughly 4,000 individuals were involved directly in nuclear bomb testing at the actual sites. And approximately 60,000 military personnel worked in the immediate area of the atmospheric tests. According to statistics published by the U.S. government, more than 99 percent of these individuals received external radiation doses of up to 5,000 millirems.

Exposure from nuclear tests conducted in the open can produce both external and internal doses of radiation. For example, among the shorter-lived radionuclides, iodine-131, with a half-life of 8 days, can still contribute importantly to both external and internal radiation exposure for months after it is formed. More durable materials like cesium-137, with a half-life of 30 years, are still with us. They were deposited in the atmosphere during the period of open-air testing and remain a source of external gamma radiation.

Internal exposure also results from radioactive fallout. As this material settles, it contaminates water, soil, and the food grown in that soil, so that radionuclides are ingested with the food. Indirect exposure can also result from what are known as food chain pathways. Fish from contaminated water and milk from cows that have eaten grass in a contaminated pasture are examples. Once inhaled or swallowed, two of the most prevalent of fallout radionuclides, strontium-90 and plutonium-239, can be expected to deliver internal radiation for an individual's entire lifetime because of their long physical and biological half-lives. Among other radionuclide components of fallout are tritium, carbon-14, iron-55, and krypton-85.

Fortunately for all concerned, all U.S. testing of nuclear devices has been carried out underground since 1963. But the fallout carried over beyond that date and did not start to fall until a year or two later. Since 1965, fallout levels have been decreasing. In 1963, for example, each individual in the United States received a total yearly, whole-body radiation dose from global fallout of about 13 millirems; by 1965 the figure was down to 6.9 millirems; in 1969 it was 4.0 millirems. As new nations join the atomic club and successfully detonate nuclear devices, the fallout level is expected to rise again, beginning with per capita doses of about 4.5 millirems in 1980, 4.6 in 1990, and 4.9 by the start of the twenty-first century. For purposes of calculating present total yearly exposure from fallout, use the average figure of 4.5 millirems per person per year.

YOUR PERSONAL RADIATION INVENTORY

Having explored the various sources of radiation in present-day America, you are now ready to calculate your own yearly radiation exposure. The accompanying inventory form has been provided to simplify this process. Anyone who has ever filed an income tax return will probably know how to complete this form.

To fill in the first blank in part A, "Natural Radiation," refer to the chart showing cosmic radiation on page 39. Locate your state or territory, read across to find the average individual dose per year, and write this number in the space provided. Thus, if you live in Alabama, you would write the number 40 on the first line.

For the next space, consult the table on page 42. Here you will find, listed by state or territory, individual dose equivalents resulting from natural radiation in the earth's crust. In the second space, then, opposite *External radiation from terrestrial sources*, the resident of Alabama would write 70.

Item 3, *internal radiation from radionuclides within the body*, is minimally dependent on geography, is fairly consistent for all individuals, and averages about 28 millirems per year. This figure has been placed in the space provided for it on the inventory form.

Now we come to part B, "Radiation Exposure Created by Human Activities." On line 4 appears the most important of these sources, radiation from medical and dental x-rays. Turn to the table on page 51. This table provides millirems of exposure to the active

PERSONAL RADIATION INVENTORY
HOW TO ESTIMATE YOUR YEARLY WHOLE-BODY
EXPOSURE TO IONIZING RADIATION

SOURCE			DOSE EQUIVALENT
A. NATURAL RADIATION			
1. Cosmic radiation			_____millirems
2. External radiation from terrestrial sources			_____millirems
3. Internal radiation from radionuclides within the body	AVERAGE		__28__ millirems
B. RADIATION EXPOSURE CREATED BY HUMAN ACTIVITIES			
4. Medical and dental x-rays			_____millirems
5. Radiopharmaceuticals			
Patient	AVERAGE	(if any)	_____millirems
Occupational	AVERAGE		_____millirems
6. Commercial nuclear power			
Environmental	AVERAGE		__1__ millirems
Occupational	AVERAGE		_____millirems
7. Industrial exposure			
Occupational	AVERAGE		_____millirems
8. Military exposure	AVERAGE		_____millirems
9. Fallout	AVERAGE		__4.5__ millirems
10. Consumer products	AVERAGE		__4.5__ millirems
11. Air travel			_____millirems
		SUBTOTAL	_____millirems
12. Any unusual exposure		(see text)	_____millirems
		TOTAL	

bone marrow; the values in the two previous tables are expressed in millirems to the whole body. This does not present so much of a problem as it may at first appear. The figures for whole-body and bone marrow exposures are obtained from similar mathematical formulas and come very close to approximating each other.

With this explanation in mind, and remembering that the figures listed in the table on page 51 are only estimates of exposure that depend on many factors (including the type and condition of the x-ray machine used for the examination and the technical expertise of the radiation technologist), simply find the procedure you have undergone, note the exposure or exposures alongside it, and put this number in the blank space opposite item 4.

Item 5 deals with radiopharmaceuticals. If you are not sure whether you received any such materials during the last 12 months, ask your doctor. If you haven't undergone any procedures involving radiopharmaceuticals, write the number 0 in the space provided. If you *have* been given radiopharmaceuticals, write in the number 3, an average figure that will probably come close to the actual quantity; radionuclides vary little in the radiation dosage they produce. Medical technicians or others who work with radiopharmaceuticals should write the average figure of 0.4 millirem in the space opposite *Radiopharmaceuticals / Occupational*.

Item 6 involves commercial nuclear power. With the number of nuclear power plants now operating in the United States, the environment inflicts on each individual a dose equivalent of 1 millirem per year, regardless of where he or she lives. This quantity has already been placed in the appropriate space. If you are involved occupationally with the generation of nuclear power, you will have to add some exposure. The average is 0.15 millirem, so you should place this figure in the blank space opposite *Commercial nuclear power / Occupational*.

Item 7 accounts for exposure you may receive if your work involves you with radiation. Examples include applying uranium glazes and manufacturing smoke alarm systems. Here again, an average figure will suffice, and the figure is a small one. Insert 0.1 millirem opposite *Industrial exposure / Occupational*.

Item 8 is military exposure to radiation. For example, the U.S. Navy uses nuclear reactors to provide propulsion for its fleet of nuclear submarines and some surface vessels. This equipment can expose naval personnel to low-level radiation while they are performing their shipboard duties. Obviously, civilian workers who repair and maintain naval nuclear reactors would also be exposed to low-level radiation. If you are in the military, or are in any way involved with military nuclear activity, write the figure 0.04 in the space alongside line 8.

As we have seen, the total yearly exposure from nuclear fallout is approximately 4.5 millirems per person. This figure has been inserted opposite item 9.

Average exposure from consumer products such as smoke alarms, luminous watches, and other devices is also about 4.5 millirems per person each year. Because this figure is roughly uniform for all Americans, it has been inserted opposite item 10.

Now we come to air travel. Systematic measurements made in flight on conventional aircraft indicate a general rate of exposure of between .25 milliren and .75 millirem per hour. Aboard the supersonic Concorde, which may travel at altitudes in excess of 60,000 feet, the exposure rate is about double that found at subsonic altitudes—.5 to 1.5 millirems per hour—but the exposure time is halved. Thus comparative exposure over a given distance is about the same for subsonic and supersonic aircraft.

To fill in item 11, estimate how many hours you spent in subsonic jet aircraft during the last 12 months. Then multiply the number of hours by .5, the average radiation exposure in millirems. If you spent 50 hours in the air, for example, you would write the number 25 on line 11. For supersonic travel, the total number of hours *equals* the average radiation exposure in millirems. A round trip from New York to Paris and back aboard the SST would take 8 hours and expose you to 8 millirems of radiation. Add this figure to your subsonic number. If you don't do any flying, you would enter nothing on line 11.

Now you are ready to add all these entries to arrive at a subtotal. For most individuals, the subtotal is also the final total. Current standards published by the Nuclear Regulatory Commission set an absolute upper limit of 500 millirems of radiation exposure per person per year and establish 170 millirems per year as the average individual exposure. These figures provide a gauge of where you stand. If your actual total is less than the recommended public safety limit of 500 millirems per year, you should have nothing to worry about. Even if your personal inventory slightly exceeds 500 millirems, do not be concerned. The normal allowable annual dose limit for radiation workers is 10 times this amount.

Feel reassured? You should—unless you're involved with exposure pertinent to line 12. *Unusual* exposure most often results from two basic sources: medical and technically enhanced natural radiation. If you or someone you know is receiving radiation therapy as part of a medical regimen, the *only* valid source of information is the physician in charge. Each individual is different, and each disease state is different. Good medical practice requires that doctors carefully consider these factors and reach decisions based solely on what is most beneficial to their patients. The risk of any exposure to radiation should always be weighed against the benefit derived from that exposure. You are certainly entitled to query any doctor about your radiation exposure. But be guided solely by what that doctor or other doctors advise you.

In the area of technically enhanced natural radiation, it is highly unlikely that you could be unaware of any such exposure. People living near the sites or former sites of uranium mining and milling operations have been the subject of more attention from the media and the federal government than most other groups. Studies are presently being conducted to assess radiation exposure and possible long-term effects.

Further, information about any radiation risk is increasingly being made available to people who live or have lived near mines, processing plants handling radioactive materials, or nuclear disposal sites. To protect the public from radiation exposure generated by these sources, federally underwritten brochures, fact sheets, pamphlets, and books are distributed to consumer groups, management groups, industrial organizations and other public interest groups. And important information about radiation is often disseminated through local radio, television, and newspapers.

KEEPING YOURSELF INFORMED

As a citizen, you have the right and responsibility to be familiar with and participate in the development of programs dealing with radiation protection. And if you are or have been exposed to technically enhanced radiation, it is even more important for you to know about the risks you may have incurred.

Many facilities are involved in radiation protection; you can obtain information by contacting the U.S. Department of Health and Human Services. Among the HHS divisions directly concerned with radiation protection and effects are the Food and Drug Administration (FDA) and its Bureau of Radiological Health (BRH), a major source of information in the radiation field. Other HHS divisions include the National Institutes of Health, the Center for Disease Control, and the National Institute for Occupational Safety and Health (NIOSH). This division conducts research activities that lead to recommendations for correcting radiation hazards in work settings.

NIOSH recommendations are forwarded for study and action to the Occupational Safety and Health Administration (OSHA), a division of the Department of Labor.

Other government departments or agencies that deal with radiation are the Environmental Protection Agency (EPA), the Department of Energy, the Department of Defense, the Nuclear Regulatory Commission (NRC), the Consumer Products Safety Commission, and the U.S. Department of Commerce Office of Radiation Programs.

For information about the radiation to which you may be exposed, write to any of them. A detailed discussion of government and state agencies involved with radiation appears in Chapter 12.

5
The Nuclear Industry

No book on radioactivity would be complete without some mention of the nuclear power industry. The demand for electric power in the United States has been increasing geometrically. A reasonable rough estimate is that our requirements are doubling every ten years. This mushrooming demand is expected to continue for the foreseeable future. With our progressively increasing need for electricity, it is almost certain that shortages will develop in conventional fuels.

Just about all the energy available to us originates either directly or indirectly from the sun. Under its life-sustaining influence, animals and plants thrive and then die. As dead animal and vegetable matter decays, and with the passage of eons of time, so-called fossil fuels such as oil, coal, natural gas, and peat develop. Of these, coal is the most abundant resource. The earth's supply of coal is probably sufficient to last several hundred years at the current rate of extraction. But oil and gas are not so widely distributed as coal; these fuels are usually found in enclosed reservoirs of definite size and volume, filling open spaces within the subterranean rock. When these sources are all tapped and exhausted, the exceedingly slow natural replacement of oil and gas will make them nonexistent for any practical purposes.

Fossil fuels pose problems over and above their shortness of supply. As these materials burn, they dirty the environment. Oxygen is consumed as carbon dioxide, sulfur dioxide, and other gases and particulate materials spew into the atmosphere. Certain weather conditions can make matters worse, trapping this noxious outpouring close to the surface of the earth, or producing corrosive sulfuric acid rain from sulfur dioxide gas. Annoying and sometimes lethal smogs that affect heavily industrialized areas and certain other geographical areas are the result.

It is obvious that alternatives to fossil fuels are absolutely necessary. Some have been tried. Solar energy, for example, has been proposed as an efficient source, without potential liability to the environment. Enormous amounts of solar energy reach the earth in the form of heat and light, but only a small part of it can be harnessed to produce energy in a form that meets our needs. Only certain areas of the world receive solar energy intense enough to use on a large scale. And, more important, production of useful energy from the sun is particularly inefficient. It has been estimated that collecting plates covering a twenty-square-mile area would be needed to gather the solar rays necessary to match one unit of energy (usually measured in kilowatt hours) produced at a conventional power station with fossil fuels.

Also complicating the use of solar energy are such technical problems as the unsuitability of elements or chemical compounds contained in the solar energy cells. Those currently used, such as silicon, cuprous sulfide, and gallium aluminum arsenide, are either too expensive, too troublesome to work with, or too difficult to produce in sufficient quantities to make their broad employment economically feasible. Furthermore, the situation would seem to call for equipment of a more advanced design than currently available so that heat gathered from the sun's rays could be transferred to another medium, such as water, with little or no loss.

Another alternative, hydroelectric plants, is much more feasible, but a limited number of suitable sites (both physically and economically) are available. The answer seems to be nuclear power, and it has been the trend in recent years. While this power source creates its own environmental problems (waste heat in the air and water and low levels of radioactivity in the atmosphere), many authorities in the field believe the trade-off of risks vs. benefits is acceptable. The alternative is to compromise our desire for a clean environment and accept a reduction in the quality of our lives.

It has been estimated that total nuclear capacity in the United States will increase from the 6,000 megawatts generated in 1970 to 800,000 megawatts by the start of the twenty-first century. In the twenty years between 1950 and 1970, while the U.S. population was increasing by about one-third, our total consumption of electrical energy more than quadrupled. Each of us, during those two decades, increased our consumption of electrical energy from 2,000 to 6,500 kilowatt hours per year. Consumption for 1970–1980 is expected to be between 11,000 and 12,000 kilowatt hours per year per person. And this statistic should more than double by the year 2000.

A great deal of this growth will almost certainly take place in the heavily populated northeastern states. According to a report to the Federal Power Commission, by 1990 it will be necessary for the power industry to develop roughly four times as much capability for generating power as it developed in the first eighty years of its existence. This report also suggests that nuclear power will account for more than 80 percent of all power generated in the Northeast by 1990. That's a lot of electricity.

THE NUCLEAR FUEL CYCLE

To achieve a better understanding of this energy source of the future, it is necessary to go back to the beginning of what has come to be known as the *nuclear fuel cycle*. Production of electricity by nuclear means involves the mining and milling of uranium—a fascinating element with an interesting history. Discovered in 1789 by a German scientist named Martin Heinrich Klaproth, it was at first a substance without a use and remained so for more than a century. A silvery metal in its natural state, it quickly combines with oxygen in the air to produce an oxide. It is in this form that uranium is contained in the ore pitchblende. Uranium is surprisingly abundant; every 250 tons of terrestrial rock contain about 1 pound. Thus uranium is about as concentrated as tin, slightly less so than lead and cobalt, and more concentrated than silver, gold, and mercury.

Before World War II, not many uses had been found for uranium, and industrial demand for it was small. The United States, which eventually became the world's largest consumer of uranium, imported the ore from abroad in those days. Most uranium arrived on our shores from mines in the interior of the Belgian Congo. Later, as demand for uranium increased due to the cold war and peacetime uses of the substance, a widespread search for domestic sources took place in the late 1940s.

Armed with Geiger counters to detect the presence of uranium, spiritual descendants of the old Forty-Niners hit the trails of the West to prospect and, they hoped, prosper. A few made millions when they sold their stakes to large mining companies. By the time the rush petered out, these concerns owned almost all of the known ore reserves in the Colorado plateau.

Uranium ore was also found in the limestone deposits near Grants, New Mexico, and in copper mines in Utah. Perhaps the most famous uranium source was the Jackpile Mine in New Mexico, from which millions of tons of high-yield uranium ore have been taken.

Mines producing uranium ore have also been operated in Alaska, Arizona, California, Idaho, Montana, Nevada, North Dakota, Oregon, South Dakota, Texas, Washington, and Wyoming. Most of the ore now mined comes from New Mexico and Wyoming. The Colorado plateau—an area that encompasses southwestern Colorado, southeastern Utah, northeastern Arizona, and northwestern New Mexico—accounts for 66 percent of the uranium oxide produced. The Wyoming basins (about 80 percent of the state) account for 23 percent, and all other regions yield about 11 percent of our uranium reserves.

In the United States today, uranium is obtained from ore by two methods. The first of these, strip or pit mining, is the most wasteful. As the name suggests, all earth that lies atop an ore deposit is simply stripped away by heavy earth-mining equipment to expose the ore. So wasteful is this process that in Wyoming, where it is the predominant method, only about 2.5 million tons of uranium ore were mined, compared to waste earth, rock, and tailings totaling 107 million tons. In Colorado, where underground mining is the preferred method, the total weight of tailings approximately equals the weight of uranium ore.

Uranium is also recovered from waste streams used by the phosphate industry. Here it is called by-product uranium. During the 1950s and 1960s, some 500 tons of uranium ore were obtained through phosphoric acid production, a method resumed in 1975. Extensive research and development have also been carried out on recovering uranium as a by-product of copper production. By the year 2000, about 25,000 tons of uranium oxide may originate through copper production. And a method was recently developed for recovering uranium from old mill tailings piles. Solution mining, as this is called, is now being used in selected locations.

But the largest source continues to be uranium mines. As of January 1, 1976, there were 17 active uranium mines in operation. They are listed in the accompanying table by company, location, and amount of ore mined daily.

Uranium Mills in the United States (as of January 1, 1976)

COMPANY	LOCATION	TONS OF ORE MINED PER DAY
Anaconda Company	Grants, New Mexico	3,000
Atlantic-Richfield	George West, Texas	1*
Atlas Corporation	Moab, Utah	1,000
Conoco-Pioneer	Falls City, Texas	1,750
Cotter Corporation	Canon City, Colorado	450
Dawn Mining Company	Ford, Washington	400
Exxon Company, USA	Powder River Basin, Wyoming	3,000
Federal-American Partners	Gas Hills, Wyoming	950
Kerr-McGee Nuclear Corp.	Grants, New Mexico	7,000
Rio Algom Corporation	LaSal, Utah	700
Union Carbide Corporation	Uravan, Colorado	1,300
Union Carbide Corporation	Gas Hills, Wyoming	1,200
United Nuclear-Homestake Partners	Grants, New Mexico	3,500
Uranium Recovery Corp.	Mulberry, Florida	2†
Utah International, Inc.	Gas Hills, Wyoming	1,200
Utah International, Inc.	Shirley Basin, Wyoming	1,800
Western Nuclear, Inc.	Jeffrey City, Wyoming	1,200
TOTAL		28,450

*Uranium obtained by solution mining.
†Uranium recovered from phosphoric acid.

Adapted from material in *Radiologic Quality of the Environment in the United States*, 1977, Office of Radiation Programs, U.S. Department of Commerce.

After uranium ore has been mined and processed, or milled, a mixture of uranium isotopes is obtained. Only uranium-235, however, is fissionable and therefore practical for use as a nuclear fuel. Thus the next step in the uranium fuel cycle is enrichment of the material obtained from milling to increase the uranium-235 content from roughly 0.7 percent to significantly higher percentages for use in nuclear reactors. We will be talking about various kinds of reactors a little later on. For reactors that use water—the most common installations—fuel is enriched to about 4 percent. Another type of reactor coming into use is the high-temperature gas reactor. Fuel for this type of reactor requires 90 percent enrichment.

The final step in the cycle is transforming rich uranium ore into fuel. This involves several chemical reactions that convert the material into a usable form. The fuel is made into pellets that are placed into tubing of stainless steel or a special alloy. The tubes are then closed off at either end and welded into what are known as *fuel rods*.

The rods are assembled in specific arrangements into fuel elements. These elements are the reactors.

FISSION

To understand the workings of a reactor, we have to go back to Chapter 3—to our discussion of how radioactivity occurs and the progression of a radioactive series. We now know that if the nucleus of a large atom is split to produce two smaller atoms, a tremendous amount of energy is released in the form of heat. The usual method of causing such a rupture of the large uranium atom is to bombard it with neutrons. Neutrons are particularly effective because they contain no electric charge and therefore are not electrostatically repelled by the nucleus, which as a whole is positively charged.

Each time a uranium-235 nucleus is struck by free neutrons and breaks apart, some new neutrons are released in the process. These in turn become available to trigger the fission of previously unaffected uranium atoms. Because it perpetuates itself and continues until all available uranium nuclei have been broken apart, this type of reaction is called a *chain reaction.*

During *each second,* more than 30 billion fissions must occur in order for 1 watt of energy to be released! But a chain reaction of this type running out of control can proceed with explosive speed and tremendous energy. The fission of one uranium nucleus, for example, produces a million times more energy than the explosion of one TNT molecule. This is the awesome power of the atomic bomb. The first of these devices that tumbled through the bomb bay of the *Enola Gay* over Hiroshima contained somewhere between 5 and 10 pounds of uranium-235, and it produced an explosion roughly equivalent to about 20,000 tons of TNT. (Subsequently, larger fission bombs used a decay product of uranium-235 called plutonium-239. Like uranium-235, this isotope can be split by free neutrons and sustains a chain reaction.)

Limitations exist on the size of fission bombs, because neither uranium-235 nor plutonium will react unless their mass is of a certain amount. This weight is called the *critical mass,* and for uranium-235 it is about 3 pounds. A lump of uranium weighing less than that is perfectly safe to store and handle. At this weight, the surface area of the uranium mass emits neutrons in such numbers that there are not

The fission process used in the atomic bomb and in all existing nuclear reactors should be distinguished from fusion, in which two atoms of a light element, such as hydrogen, are joined together (fused) into one larger atom, again with the production of gigantic amounts of energy. Reactions of this type produce the energy of the sun and stars, particularly from the fusion of hydrogen nuclei to form helium. This is the principle of the hydrogen bomb. Indeed, the sun is nothing more than a massive hydrogen bomb fusing hydrogen atoms together constantly at virtually unmeasurable temperatures.

The fusion of hydrogen nuclei to form helium takes place in a series of stages that produce deuterium (heavy hydrogen) and tritium (hydrogen-3). And with each of these reactions, energy is released. But none of these steps can occur unless temperatures in the millions of degrees are reached. High temperatures cause the light particles to move at the enormous speeds required for them to get close enough together to fuse.

The hydrogen bomb tested at various Pacific Ocean atolls in the period after World War II consisted of a uranium-235 or plutonium fission bomb of the type dropped on Hiroshima and Nagasaki, surrounded by a jacket of hydrogen. When the fission bomb is exploded, it produces the free neutrons and the high temperatures that give them enough energy to fuse with the hydrogen nuclei. Reactions that occur at these tremendous temperatures are called thermonuclear reactions. Materials in a fusion reaction must be heated to temperatures of between 100,000,000 and 1,000,000,000 degrees—higher than the temperature at the center of the sun. This heated gas mixture, or "plasma," must then be contained long enough for the reaction to occur.

What makes thermonuclear weapons even more scary is the lack of limitation on their size. Unlike fission bombs, hydrogen weapons can be made to virtually any size or destructive potential. Whereas the first atomic bomb was equivalent to 20,000 tons of TNT, hydrogen bombs can be made a thousand times more powerful. They can explode with a force equal to 20 million tons of TNT. In fact, thermonuclear weapons in excess of 50 million tons have been tested.

Controlled fusion reactions are not presently feasible, given the current state of our technology. The use of nuclear fusion for peaceful purposes will not occur for many years to come—probably not until the twenty-first century at least. This is unfortunate, because fusion reactions offer important advantages over fission reactions. As a raw material, fusion reactions use the abundant element hydrogen (every molecule of water contains two atoms of hydrogen) instead of uranium. And no radioactivity is involved. However, until scientists solve the problem of developing a container to house a reaction taking place at higher temperatures, there will be no fusion reactors. For the time being, generation of nuclear power depends on the use of radioactive materials.

cnough of them within the mass to keep a chain reaction going. If the mass were increased by degrees, however, its weight would become greater in proportion to its surface area. When the critical mass was reached, there would be enough neutrons within to sustain a chain reaction. (The atomic bomb was actually made of two masses of uranium-235, each weighing less than the critical mass but more than half that weight. Bringing these two masses together very rapidly created one lump heavier than the critical mass, got the uncontrolled chain reaction going, and caused the explosion.)

THE NUCLEAR REACTOR

While it may be fine for bombs, an uncontrolled chain reaction will not do for the commercial production of energy. The reaction must be slowed down and controlled. For this purpose, an atomic pile or (as it is now called) a nuclear reactor is needed. Its function is to furnish a particular set of conditions in which fission reactions can be started, kept going, and controlled. And it must provide a means of recovering the heat generated by the reaction.

At the heart of reactor control is the need to maintain a steady power level. In other words, the neutrons released with each fission must be slowed down so that they cause only one more fission to take place, rather than letting the chain reaction "run away," or explode. The method used to slow neutrons down is to put obstacles in their path; you can run in an open area, but on a crowded city sidewalk you must proceed slowly. Therefore, the free neutrons are allowed to collide repeatedly with light nuclei of other elements that will not themselves capture the free neutrons or split apart. Substances that perform this function are called *moderators*, and they are usually mixed with the uranium. Carbon, or graphite, and heavy water are most often used for this purpose. Most modern reactors use heavy water.

Another important means of keeping the chain reaction in hand is the use of *control rods*. While the *speed* of the neutrons is controlled by the moderators, the actual *number* of neutrons on the loose in the reactor is regulated by the control rods. Unlike the moderator, control rods capture free neutrons rather than simply deflecting them. For this purpose, the metal cadmium seems to function ideally. Cadmium or cadmium-plated rods are inserted into the chamber that houses the nuclear reaction, and fission begins.

As neutrons are liberated, they are slowed down by the moderators and absorbed by the control rods. When these rods are kept in the fully lowered position, a chain reaction cannot take place. As they are retracted, less and less cadmium remains inside the reactor to absorb neutrons. When the control rods are withdrawn to exactly the level that permits absorption of just enough neutrons to prevent an explosion, but not quite enough to snuff out the chain reaction, the fission reaction is said to be balanced. Provisions exist for the rods to be fully lowered automatically, stopping the reaction, if it should show signs of running away.

Two additional elements are needed for a nuclear reactor. One of these, the *cooling fluid,* can also act as the moderator if heavy water is being used for this purpose. The heat that the nuclear reaction generates is collected by the cooling fluid within the reaction chamber. Because of the neutron activity taking place in the chamber, this fluid becomes radioactive and is made to recirculate within the system to sidestep the problem of disposal. Via this closed circuit, the heat of the heavy water is transferred to ordinary water by means of a device called a heat exchanger.

Finally, a nuclear reactor must provide a *shield* to protect life from the dangerous, high-energy gamma rays liberated during the nuclear

DESIGN OF THREE MILE ISLAND REACTOR

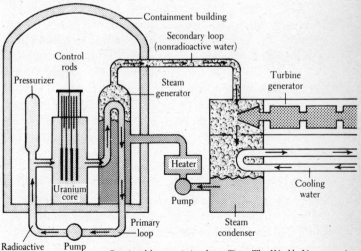

Reprinted by permission from *Time*, The Weekly Newsmagazine; Copyright Time Inc. 1979.

reaction. For this reason, most reactors are encased in jackets of concrete several feet thick. An efficient absorber of both gamma rays and neutrons, concrete performs its shielding function most efficiently.

What are such systems used for? Reactors can be applied to a variety of uses. First, of course, is scientific research in such areas as nuclear physics, radiation chemistry, analytical chemistry, biology, and medicine. Reactors used for these purposes are generally designed so that the reactor core sits in an open pool of water, permitting it to be shifted to accommodate special experimental apparatus.

However, the widest use of nuclear reactors is in the generation of electricity, and their application in this context is really very simple. In an atomic power plant, the nuclear reactor merely performs the function of a boiler in a conventional installation. Ordinary steam-electric power plants burn fossil fuels (coal, oil, or natural gas) to produce heat and steam. The steam is then harnessed to turn the turbogenerator that produces the electricity. In a nuclear power plant, the required heat is produced in a nuclear reactor. Transferred from the cooling fluid to ordinary water by means of the heat exchanger, this heat produces the steam that drives the turbogenerator that produces the electricity. Thus a highly sophisticated, state-of-the-art nuclear reactor turns out to be nothing more than a fancy substitute for the old-fashioned boiler.

The first nuclear reactor plant for the production of electricity was built in the Soviet Union. It was a small power station with a capacity of 50,000 kilowatts, and it went into operation in June 1954. Great Britain was next in the nuclear power parade, commissioning Calder Hall in October of 1956. This installation had a capacity of 90,000 kilowatts. On May 26, 1958, the United States joined the club with the Westinghouse nuclear reactor at Shippingpoint, Pennsylvania. The facility was capable of generating 60,000 kilowatts. Then, in 1960, a 180,000-kilowatt reactor began generating electricity in Illinois, and a power station with a 110,000-kilowatt capacity opened in Massachusetts.

Over the next several years, many more installations were commissioned. By 1974, 44 civilian nuclear power reactors were operating in 18 states. And today the number of reactors is pushing toward 80, with almost 130 new plants in the planning or building stages. The accompanying map, which is based on information available from the

Department of Energy, shows the locations of plants operating, under construction, or planned as of January 1, 1979.

VARIETIES OF REACTORS

In the United States, the majority of our nuclear reactors are the light-water variety, in which ordinary water both moderates the neutrons produced by the nuclear reaction and serves to transfer the heat of the reaction. Two basic types of these reactors exist. In the pressurized water reactor, the water does not boil. Instead, it is pumped under pressure at high temperature through a heat-exchanging device that transfers the heat to water in a secondary circuit, which boils and forms steam. The steam, of course, drives a conventional turbine that runs the generator. The second kind of light-water reactor uses boiling water. Inside the reactor, the water is allowed to boil and form steam, which runs through the turbine directly.

In both types of water-cooled reaction, the steel containers that house the radioactive materials are constructed to very stringent standards and must be capable of withstanding extremely high temperatures and thousands of pounds of pressure per square inch. These are not particularly efficient methods of turning heat into electricity; only about one-third of the heat is converted. In both of these

NUCLEAR REACTOR LOCATIONS

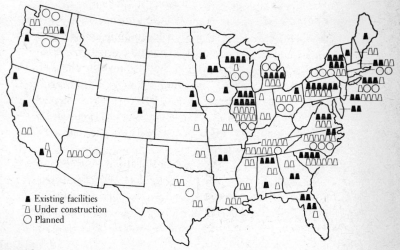

▲ Existing facilities
⌂ Under construction
○ Planned

systems, the primary cooling fluid is made to circulate in a closed circuit, completely removed from the lake, river, or ocean where it originated. This is not unusual. In all U.S. facilities, the only water drawn from a natural source that flows through the system and returns to its source is the water used to cool the turbine condensers. This water never enters the reactor; its only function is to dissipate nonusable heat.

Occasionally, the primary cooling fluid undergoes purification to limit its level of radioactivity. The purified water is then returned to the system. Purification is also carried out on other fluids in the plant —water used to wash equipment as part of plant maintenance, for example.

A major ecological concern is the condition of waterways near nuclear power plants. Do any radioactive materials find their way into these streams? The answer is yes. Most of the solid or liquid radioactive materials removed during purification are stored for future disposal. We'll have more to say about this "nuclear garbage" a little later on. However, an infinitesimal amount (perhaps only a few millionths of a gram each day) is flushed into the waterway serving the station. This stream is heavily diluted and is hardly radioactive at all. In fact, it can pass federal standards for drinking water. And the radioactivity level drops even further as the waste stream is diluted in the natural water source.

There also exists a reactor that creates no radioactive water. This is the high-temperature gas reactor, in which the neutrons are moderated by solid graphite and heat is removed by means of helium gas at high pressure. After being heated to high temperatures, this gas passes through a heat exchanger where heat is transferred to ordinary water, which boils, forms steam, and drives a turbine. This water flows in a separate system and never makes contact with radioactive material.

In addition to the light-water reactors are those that use heavy water to channel heat away from the reactor. The advantage of this type of reactor is that it doesn't require enriched uranium-235, as the light-water reactors do. And there's a good reason.

Remember that only 1 of every 140 uranium atoms found in nature is uranium-235, the fissionable uranium isotope. The more common isotope is uranium-238. This substance isn't split when struck by a neutron, but it will absorb that neutron and emerge as uranium-239.

By emitting 2 beta particles, the molecule decays to neptunium-239 and then to plutonium-239, a fissionable isotope. Heavy water, of course, both supplies the spare neutron that sustains the chain reaction and converts uranium-238 to uranium-239.

Because this type of reaction produces more fissionable substance than it consumes, it actually "breeds" fuel. Reactors that work on this principle are known as *breeder reactors,* and this type of reactor can also use thorium as fuel. When the naturally occurring isotope, thorium-232, is peppered with neutrons, uranium-233 is formed. This isotope is also fissionable. Thorium-232 occurs abundantly in nature, and low-grade deposits have been found in large concentrations along the western reaches of the Appalachians in eastern Tennessee and in areas of Kentucky, Ohio, Indiana, and Illinois.

The latest word in breeder reactors is the so-called liquid-metal fast-breeder reactor. This type of device incorporates both plutonium and uranium into a core immersed in a large reservoir of liquid sodium. This liquified metal flows through the reactor, absorbs the heat of the nuclear reaction, and carries it to the essential heat exchanger, which heats water in a separate closed system to produce the steam that powers the turbines.

Thus we see that all nuclear reactors are not equal in their consumption of nuclear fuel. Most of the nuclear power plants operating in the United States today are profligate consumers of fuel. Somewhere in the middle of the scale are reactors with a more modest rate of fuel consumption. And finally there are the breeder reactors; their fuel consumption is negative because they create more fuel than they use. When breeder reactors become feasible for commercial use, experts estimate that low-grade deposits of uranium-238 and thorium-232, mined cheaply from shale, will provide prodigious quantities of fissionable material—enough to meet the energy needs of the United States for 100 millennia or more!

THE THREAT OF ACCIDENTS

Until breeder reactors become feasible, used fuel from conventional nuclear power plants must be reprocessed to recover plutonium and uranium isotopes. This somewhat complicated process results in large amounts of radioactive waste material. It goes without saying that the handling and disposal of this material at a reprocessing plant are critical concerns, as is the discharge of low-level radioactive waste into the environment. Of the gaseous and liquid wastes that are generated through fuel reprocessing, the most common are carbon-14, krypton-85, iodine-129, tritium, rubidium-106, strontium-90, cesium-134, cesium-137, uranium, and plutonium. Careful monitoring of these wastes is essential.

The future of reprocessing operations is undecided. Actually, no plants have operated in the United States since the early 1970s. A reprocessing plant in West Valley, New York, operated from 1966 to the end of 1971, but was shut down so that it could be redesigned and modified to increase its reprocessing capability. These plans were abandoned in 1976 for economic reasons.

In the mid-1970s, construction on another reprocessing plant was begun in Barnwell, South Carolina. As of this writing, the plant awaits licensing and has not yet begun to operate. Finally there is the General Electric Nuclear Fuel Plant in Morris, Illinois. Here technical difficulties have made it unlikely that the facility will ever operate. In the meantime, spent fuel continues to be stored in special areas at individual nuclear power plants and in storage areas in West Valley and Morris.

TRANSPORTATION OF RADIOACTIVE MATERIALS

How spent fuel gets to these storage facilities is another major source of concern. What risks to the general population arise from the routine transportation of radioactive materials through communities and from possible accidents involving vehicles carrying such materials?

A study on this subject was done for the Environmental Protection Agency in November of 1974. Among the materials considered were spent fuel, plutonium, high-level radioactive solid waste, and gases resulting from fission. Transportation modes included truck, rail, and barge. This study found that the probability of accidents is low; that when accidents do occur, only a small number of them involve the release of radioactivity; and that, in the majority of these cases, only a very small fraction of the radioactive content is released. For these reasons, the EPA study concluded that the public health risk involved in transporting radioactive materials is relatively small.

In an accident, the person most likely to be injured is the driver. And the chances are thousands of times greater that his injury will result from the crash itself than from radiation.

In the United States during the last quarter-century, about 300 accidents involving the transportation of radioactive materials have been reported. And approximately 30 percent of these accidents resulted in the release of radioactive material from medical and industrial radiochemicals, not from wastes that originated at a nuclear power plant. None of the accidents caused death or injury as a result of the radioactive nature of the material. No release of radioactivity has occurred from shipments originating at nuclear power plants. Danger to the public would thus seem to be negligible compared to that involved in transporting gasoline, naphthalene, liquefied natural gas, and other inflammable or explosive products.

In one accident, a tractor-trailer carrying radioactive waste from a power station in Illinois to a disposal site in South Carolina collided with another vehicle. Drums containing the radioactive material were spilled onto the highway, but these containers did not rupture. Nevertheless, many towns do not permit transportation of nuclear materials within their limits.

PLANT SAFETY

While many Americans are worried about transportation of radioactive materials, a far greater number are increasingly disturbed by the environmental effects of nuclear plants and the possibility of disaster due to nuclear accidents at the plants themselves.

Prior to Three Mile Island, no nuclear accidents affecting the general population had occurred in any nuclear power facility in the United States. To be sure, routine operation of a nuclear power reactor releases tiny amounts of radioactivity into the environment. These levels are tolerable by most criteria, but some members of the scientific community believe that *no* dose is safe. In response to critics of nuclear power, the U.S. Atomic Energy Commission introduced a rule limiting the dose at the boundary fence of a nuclear reactor. An individual who stood on that spot for a year would receive a maximum dose of only 5 millirems.

This is the amount of radioactivity one would absorb by taking a jet across the Atlantic once or by living in a concrete building a few weeks each year. The critics of nuclear energy tend to ignore the greater hazard to the public arising from ordinary coal-burning power stations. Clouds of heavy black smoke continually belch from these plants, and this noxious miasma contains poisonous gases, soot, and dust particles—pollutants that probably cause significantly more harmful effects than routine emissions from nuclear reactors.

We should note that, since nuclear reactors came into being, a few individuals have lost their lives while working in these plants. But the entire nuclear power industry can boast an excellent safety record. Very few lives have been lost due to high doses of radiation attributable to activities at or related to nuclear plants. Statistically, it is far more dangerous to work in an ordinary factory than in a nuclear reprocessing or reactor plant.

If Three Mile Island did nothing else, it put the word "meltdown" into the average American's vocabulary. The term refers to what happens when a nuclear reaction gets out of hand and generates so much heat that it dissolves not only the core but also the nuclear reactor itself, the foundation on which the structure stands, and finally the very earth into which it melts. Theoretically, the runaway reaction would continue on its fiery path through the earth's core and completely pass through the planet to emerge, perhaps, in China— the so-called China syndrome. For present reactors, both water- and gas-cooled, a meltdown produced by loss of coolant is termed the maximum credible accident. And the Three Mile Island facility came perilously close to just this condition when its emergency cooling system was inadvertently shut off.

Most nuclear reactors are engineered so that emergency cooling systems cut in when they are needed. But are such systems really adequate? Unless they provided enough coolant water in a very short time, the fuel in the reactor core certainly would melt, and the molten mass would be certain to break through the containment chamber. In such an event, large quantities of gaseous, volatile, and nonvolatile products of fission would be released into the atmosphere as a radioactive cloud, exposing the population downwind from the reactor site to considerable radiation. Radioactive steam was released at Three Mile Island, but the amount was small and its adverse effects seem to have been minimal. This may have been because authorities ordered the evacuation of all pregnant women and preschool children within five miles of the plant site and advised everybody to stay indoors.

Three Mile Island was the worst commercial nuclear accident in the history of the United States. But there have been other near-calamities. In January 1961 at a test site in Idaho Falls, an experimental reactor exploded, killing three technicians. Radiation levels inside the facility were so high that it was necessary to bury exposed parts of the victims' bodies with nuclear waste material. Fortunately, no radiation escaped from the plant. In investigating the accident, the Atomic Energy Commission concluded that a control rod might have accidentally been withdrawn too far. (In a bizarre footnote to this incident, a former AEC official circulated a memo alleging that one of the victims may have deliberately sabotaged the plant as a means of avenging himself on a fellow technician whom he suspected of having a love affair with his wife.)

At the Enrico Fermi facility, an experimental breeder reactor situated a few miles from Detroit, the core partially melted down when a cooling system malfunctioned. A back-up system successfully checked the out-of-control reaction. However, in something of a cover-up, officials did not disclose fears that further melting of the fuel would take place. Evacuation of the 1.5 million residents of Detroit was even considered at one point. So high were radiation levels inside the Detroit Edison facility that it could not be entered for a month, and it was four years before the reactor went back into operation. In 1972, this plant's license was revoked and it was decommissioned.

Water levels rose and fell erratically when a malfunction triggered mechanisms that closed down the Commonwealth Edison Dresden II reactor in Morris, Illinois, during June of 1970. Human error compounded the problem when a broken gauge was not recognized as such. More water was pumped into the system, and the resulting increase in pressure forced 50,000 gallons of radioactive fluid out of the reactor vessel and into the containment area.

In March of 1975, a bad fire was started at the Tennessee Valley Authority Brown's Ferry reactor in Alabama when a technician went searching for an air leak with an open flame. The fire destroyed the wiring that led to the reactor controls. With no controls operating, pressure rose sharply in the reactor, causing the cooling system to fail as the water level dipped precipitously. Ten hours elapsed before equipment could be repaired and a water pump put in operation to avert a potential meltdown.

In Richland, Washington, during November of 1977, the Hanford reactor leaked 60,000 gallons of radioactive water into the Columbia River, and activities at the plant were temporarily suspended. According to officials, radiation levels were not high enough to harm humans or animals. This plant seemed to be jinxed; only a month later fuel leaked into an elevator, slightly contaminating four workers.

For two hours during a cold day in January of 1978, radioactive helium spewed from a chimney at the Fort St. Vrain nuclear facility in the Denver area. Officials reacted with commendable speed, evacuating 200 plant personnel, alerting hospitals, and sealing off all access roads within a 5-mile radius. Radiation at the Public Service Company of Colorado plant shot up to 30 times its normal level but only lightly contaminated a few employees.

Then, as a new decade dawned, a new area of concern arose. The previous near-misses could be chalked up to human error or to the malfunctioning of equipment. But in January of 1980, a minor to moderate earthquake occurred in the San Francisco Bay area, disturbing footings of the nuclear reactor at the Lawrence Livermore Laboratory. A small radiation leak that developed was quickly stopped and caused no problems. But the incident dramatically underscored the possibility of nuclear disaster through natural catastrophes.

These accidents were certainly not lost on the great many people who have never believed that nuclear energy is safe. Nevertheless, in the first twenty years of the nuclear power era, not one death of a person working outside the nuclear industry has resulted from a nuclear accident. Nuclear industry spokesmen point out that we don't ban automobiles, which cause 50,000 deaths each year. Why, then, should we abandon an industry that has saved the United States approximately $2 billion (that would otherwise have been spent on imported oil) solely because of an accident that killed no one?

The logic is flawless, but many knowledgeable people are not convinced that a calamity could not occur. While these opponents concede that the odds against such an occurrence are hundreds of thousands to one, the findings of the commission empowered to investigate the accident at Three Mile Island are not reassuring. A year after the near-disaster, this group could offer no unqualified guarantees that serious nuclear accidents would not occur in the

When we speak of nuclear accidents, we are not talking about nuclear plants exploding in mushroom-shaped clouds. There is universal agreement among authorities that light-water reactors cannot explode like an atomic bomb. The result of a major nuclear accident is simply the release of radioactivity. But the consequences of such an event are obviously much greater than those of floods, tornadoes, forest fires, or earthquakes. While thousands of people may be killed and hundreds of millions of dollars lost in damage, these non-nuclear catastrophic events are self-limiting. There are no long-range effects, no genetic consequences, and no large areas of land made unfit for human habitation for years.

future. In a 179-page report, the twelve members of the commission found serious deficiencies in training of personnel, procedures followed when an accident occurs, and design of nuclear equipment.

The primary cause of the accident at Three Mile Island turned out to be human error, and even the most sophisticated equipment cannot prevent people from making mistakes. All nuclear reactors are equipped with elaborate back-up systems, and these back-up systems are equipped with a back-up system of their own. Thus, if a cooling system fails, two safety systems are immediately activated. One plunges control rods into the core, and the control rods shut down the reaction by absorbing neutrons. Another—the emergency core-cooling system, or ECCS—floods the reactor with water from an alternative source. But at Three Mile Island, a technician actually turned off the ECCS. If this back-up system had been kept on during the early stages of the accident, the entire affair would have been a relatively minor incident.

In another much-discussed incident at Millstone One, a reactor near New London, Connecticut, a technician inserted control rods in the wrong order following a fuel replacement operation. Two of the rods touched, and fission began with the top of the containment vessel open. Here, a back-up system did its job properly. Automatic controls immediately stopped the reaction before any radiation could escape.

Human error may well be the one factor that causes critics of nuclear energy the most concern. According to the Three Mile Island commission, technicians at the plant were so conditioned to think in terms of averting large-scale accidents that they were not aware of how to deal with the small, less dramatic equipment failures that caused the accident they were confronted with. The personnel charged with containing the accident became confused. Thus a relatively insignificant incident burgeoned into a major episode that might have done great harm to the populace and did millions of dollars worth of damage to the reactor.

In fairness to the people who worked at Three Mile Island, we should re-emphasize that the fault was not all theirs. Some of the blame must be laid at the door of the engineers and scientists who designed the control room panels and equipment. For some inexplicable reason, the conventional color codes of red for "stop" or "danger," and green for "go" or "safe," were reversed in certain areas of the control room.

Further evidence of poor design appears in the distance (as much as 80 feet) between meters and related switches, a fact that caused control panel operators to claim they needed roller skates to do their jobs properly.

Faulty design is also evident in reactor control-room panels, including those at Three Mile Island. The panels bristle with row upon row of indistinguishable dials and handles identified only by small, barely legible labels. Rarely used meters and switches—the devices needed only during emergencies—are hard to find, because the designers place frequently used controls in the more convenient locations. A technician would therefore be relatively unfamiliar with the emergency controls and could become confused in a moment of stress. Obviously, control knobs for similar functions should be similarly shaped, and those for different functions should include clear, distinguishing features to help minimize the possibility of operator error.

Critics of nuclear power also indict plant design as a potential cause of accidents. And mistakes in design do occur. A week after the Three Mile Island accident, for example, officials at the Georgia Power Company confirmed that design problems had been encountered during testing of a new plant. A 40-foot length of pipe in the reactor cooling system broke loose, and when the damaged pipe was inspected, officials discovered that its bolts had been faultily installed.

In California, various protest groups delayed for 10 years the opening of a Pacific Gas and Electric reactor, the 2,000-megawatt Diablo Canyon facility located between San Francisco and Los Angeles.

Final licensing approval from the Nuclear Regulatory Commission was also seriously delayed. The problem stemmed from a geology report revealing that a major earthquake fault lay just 2.5 miles from the plant site. The two reactors at Diablo Canyon were constructed to withstand an earthquake registering 6.5 on the Richter scale, but the nearby geological fault is thought capable of producing quakes in the 7.5 range. The plant has been heavily reinforced and stabilized, but doubts still exist.

PLANT SECURITY

Among other major questions raised about plant safety and the possibility of a nuclear accident is plant security. A constant worry among those who oppose nuclear power *and* those who support it is the possibility of terrorism. Many studies confirm that a well-trained individual or group could sabotage a plant or capture one and hold it for ransom. The likelihood of a nuclear accident could increase significantly during such terrorist activities.

Most nuclear power plants, of course, are equipped with elaborate security systems. So sensitive are their detection systems that a large bird landing on one of the fences that surround the plant can set them off. Operators peer from their control rooms through bullet-proof glass, and most installations teem with heavily armed special police.

Some of the intruder-detection devices being tested sound like science fiction. For example, a laboratory in Albuquerque is testing a device that responds to the presence of an intruder by filling the room with a sticky, taffy-like substance that traps and holds the trespasser in much the same manner in which flypaper captures a fly.

THE DISPOSAL OF NUCLEAR WASTES

The safe management of nuclear garbage, or waste materials, will also be pivotal in the future of nuclear power. At this point, we should draw a distinction between the terms "effluent" and "waste." Effluents are by-products of the nuclear fuel cycle released into the atmosphere, into surface streams, lakes or oceans, or onto the earth itself. Nuclear waste, on the other hand, is material that is kept for the duration of its hazardous lifetime under conditions designed to isolate it from living organisms. This usually involves burial or some other means of isolation to ensure that the material will not escape into the environment.

DIFFERENT WASTES, DIFFERENT PROBLEMS

The Environmental Protection Agency (EPA) is charged with managing nuclear waste, and it does not differentiate between sources. The agency takes the view that radioactive waste in any shape or form constitutes a potential health and environmental hazard. Waste is waste, whether it comes from a nuclear reactor, a laboratory, or any other facility where radioactive material is handled.

All radioactive waste, however, is not created equal. Some waste materials take up more space than others; wastes may vary in chemical composition; and requirements for treatment or handling may differ from material to material. Accordingly, the EPA has classified radioactive wastes into six general categories: high-level; low-level; contaminated with transuranic elements (those into which uranium decays, such as plutonium); uranium mill tailings; any and all materials at a radioactive facility or area that has been taken out of operation; and naturally occurring radioactive substances produced by other than nuclear-energy industries, such as coal mining and phosphate processing.

The U.S. Nuclear Regulatory Commission (NRC) has also classified nuclear waste. Their categories include high-level; low-level; wastes containing transuranic elements; contaminated facilities and equipment; and spent fuel not slated for recycling.

A word about some of these classifications is in order. High-level waste is that material which remains after the reusable fuel has been separated from spent fuel in a nuclear reactor facility. Low-level waste, on the other hand, comprises a great many industrial materials that become contaminated with radioactivity in the course of their handling. Low-level waste usually occupies a much greater space than high-level waste, but it contains significantly less fissionable material and remains radioactive for relatively short periods of time.

Low-level wastes that are not transuranic (that is, they are free of radionuclides resulting from uranium decay) usually contain isotopes with maximum half-lives of roughly thirty years or less. Given these relatively short life expectancies, all of this material will decay to completely harmless levels in just hundreds of years. By contrast, the transuranic elements in high-level waste material may have half-lives numbering in the tens of thousands of years. Obviously, this degree of longevity must be taken into consideration when disposing of transuranic waste, whereas the requirements for low-level material need not be so stringent.

In other words, the long-term hazards associated with low-level and high-level waste are not necessarily proportional to the level of radioactivity, but rather to the decay rates of the radionuclides. And the specific radionuclides found in high- and low-level waste decay with half-lives ranging from a few months to hundreds of millennia.

As for facilities and equipment, they themselves become radioactive waste material after being taken out of service, or decommissioned. How this is done depends on the type of facility, but the equipment must be removed and all surfaces checked for potentially dangerous radioactive material. That which exists, of course, must be disposed of.

One of the many major problems confronting authorities since the dawn of the nuclear age is what to do with spent fuel. So far, the most workable solution has been to bury used fuel permanently in stainless steel containers thousands of feet underground. The only problem is finding a suitable location.

Ideally, radioactive material should be buried in a site free of geological faults, so that no seismic activity can jar the containers and cause them to break. The site should also be relatively dry, so that containers will not corrode. Ground water that rusts a containment vessel and causes it to leak radiation can itself carry radioactivity into the water supply. Excavation into salt deposits and tunnels dug into mountains of granite and basalt formations would seem to offer the best possibilities.

Scientists also talk of a system of tunnels and container tanks with life expectancies of at least a hundred years, plus monitoring systems that can expedite the shunting of radioactive waste from a damaged tank to an intact one without any loss of containment.

Considering these physical requirements, the most likely region for locating nuclear waste repositories is in the western part of the United States. However, just as some communities will not permit transportation of nuclear wastes within their boundaries, many states are denying access to their lands as dumping grounds for nuclear garbage. In the meantime, about 10 million pounds of used fuel from the nation's nuclear reactors are being stored in huge underwater tanks at or near the reactor locations. Here, apparently, the waste will stay until a permanent burial ground is found where the spent fuel can be interred as long as is necessary to render it harmless.

Of all types of nuclear waste, liquids containing high levels of radioactivity are the most difficult to deal with and pose the most severe potential hazards. During the past thirty years or so, federal agencies involved in research and development for both military and peaceful uses of atomic energy have produced more than 205 million gallons of high-level radioactive liquid waste. Most of this lethal broth is stored at three sites: the Hanford Reservation in Richland, Washington; the Savannah River facility near Aiken, South Carolina; and the Idaho National Engineering Laboratory near Idaho Falls, Idaho. Richland accounts for a little over 75 percent of the total, Savannah River for about 23 percent, and Idaho Falls for 2 percent to 3 percent.

This material, of course, is of the high-level variety. And contained radionuclides that cause the greatest concern are strontium-90, cesium-137, and plutonium-239. These radionuclides cannot be neutralized, and each is particularly dangerous because of the damage it can do to the human body.

For example, strontium-90, an emitter of beta particles, and cesium-139, a gamma-ray emitter, require about 600 years to decay to one-millionth of their original radioactivity levels. To reach the same level, plutonium-239, an alpha-particle emitter, requires about 500,000 years. These three radionuclides not only pose problems for long periods of time, but they can also enter the human body insidiously in a variety of ways—via the air we breathe, water, vegetables, and meat, and animal products such as milk and cheese.

To minimize the possibility that any of these three villains will get loose, the radioactive liquids being stored are solidified to prevent leaks and reduce their volume. Waste management is also aimed at developing ways to keep the solid waste from migrating to where it is not wanted. How this is accomplished varies from site to site and depends on such factors as the geology of the area, the prevailing weather, and the composition of the waste material. At Richland and Savannah River, the high-level liquid waste is fashioned into something called salt cake; at Idaho Falls, it is converted into a dry, granular form known as calcine. Neither of these solids is the definitive answer for long-term storage, because they can be broken up by the action of water, but no other form has yet proved commercially feasible.

HOW MUCH IS THERE?

Solutions to this and many other problems will have to be found. Radioactive debris is just not going to disappear. In fact, it is estimated that 60 million gallons of high-level "civilian" waste will be produced by the year 2000. Twenty years later, the figure is expected to be 238 million gallons! All this, of course, is over and above the 205 million gallons generated during the last three decades or so, and the roughly 7.5 million gallons generated by federal agencies every year.

Just how much of a hazard is involved in storing such huge quantities of liquid radioactive waste? Figures covering the period up to June 1974 indicate that 26 leaks had occurred in underground storage tanks. Eighteen of these occurred at Richland, where 430,000 gallons of waste seeped into the ground. The remaining leaks occurred at Savannah River, but only one of them released waste into the ground. To date, the Idaho Falls installation has not experienced any leaks. According to the Department of Energy, there was no contamination of the environment except the ground beneath the tank. Further, there was no exposure to the public as a result of leakage, because the radioactive matter was immobilized in the soil.

Of course, the possibility of leakage is the major reason for solidifying the liquid to salt cake or calcine form. Through this program, the total volume of high-level liquid waste at the three storage sites operated by the federal government declined from 92.3 million gallons at the end of 1967 to 80.8 million gallons in June of 1974.

In the private sector, approximately 600,000 gallons of high-level liquid waste have been produced, and this material is stored on state-owned land at West Valley, New York. According to a regulation issued by the government in 1970, commercial concerns that produce high-level liquid waste are given five years in which to convert it into an acceptable solid form. Excluded from this edict are the 600,000 gallons at West Valley. The solid waste is then turned over to the Department of Energy no more than ten years after it was produced. From that point on, the government agency assumes physical custody of the waste matter for permanent storage, with the cost passed on to the commercial facility that generated the waste. Research is presently underway to develop the technology that will permit ultimate disposal. It is of utmost importance to all of us that these efforts succeed.

Experts have estimated that, if the total volume of solidified, high-level waste material generated by the commercial nuclear industry by the turn of the century were formed into a cube, each side would be 70 feet In length.

WHAT DO WE DO WITH IT?

There is no dearth of ideas about how to dispose of high-level wastes. In addition to the methods we have mentioned, other possibilities include shallow burial in desert areas, disposal in extremely deep shafts drilled into rock, dumping into rock reservoirs formed by radioactive melting, disposal of liquid in subterranean wells, insertion of the waste deep under the floors of oceans, burial under ice in Antarctica, and propelling the material by rocket into outer space. Most of these methods are somewhat visionary and must await much more research activity.

In the United States, the preferred method of radioactive waste disposal has been burial in salt beds. But this practice has recently been discontinued in favor of carefully engineered storage facilities with specially designed vaults located on the earth's surface. Before the radioactive material is placed in the vaults, it is first sealed into an epoxy material or embedded in concrete much like the victims of gangland killings. The above-ground disposal method offers certain advantages. First, the material can be checked for leakage periodically, and the appropriate steps taken if any leak is discovered. Second, material stored in this way can be re-disposed of if scientific research yields new and superior methods.

For the disposal of low-level wastes, the government has used three methods: diluting the material and dispersing it widely, disposing of it at sea, and burying it in shallow pits in the earth. Remember that the isotopes of concern in low-level waste are relatively short-lived; this material need not be isolated for so long a time as high-level waste.

Of the three methods, dilution/dispersion has fallen into disfavor. Releasing greatly diluted radioactive waste into natural waterways is still permissible under existing regulations, but pressure is being brought to bear to keep such contamination of the environment to the lowest possible levels. Apparently there are no radiation Love Canals, only chemical ones.

In June of 1970, the same concern for the environment resulted in the discontinuation of low-level waste disposal at sea. However, during the twenty-four years between 1946 and 1970, large quantities of low-level package wastes were thrown into the oceans. The bulk of this material was generated by research and development facilities rather than commercial plants. Fortunately, most of the radionuclides in the waste material were short-lived and have long since decayed into innocuous substances. This is a far cry from a contamination incident that occurred in Bloomfield, Colorado, in 1973. Tritium was inadvertently released from a nuclear plant into the Great Western Reservoir, which supplies drinking water to approximately 13,000 people in the area. As a result of this incident, authorities estimated that each resident of Bloomfield who consumed an average quantity of water each year received approximately 6 millirems of radioactivity.

Regulatory authority to issue permits for ocean disposal of radioactive waste rests with the Environmental Protection Agency. However, the agency has issued no permits since its inception in 1970. During 1974 and 1975, the agency examined the no-longer-utilized dump areas in deep ocean waters to check on the status of materials that had been jettisoned there, and to determine whether any waste had seeped into the sea. The sites visited included areas in the Atlantic Ocean (off the coasts of Maryland and Delaware) and in the Pacific Ocean (not far from the Farallon Islands). These regions had received the lion's share of all material dumped at sea from 1946 to 1962, although the Gulf of Mexico was also used as a dump site. Everything seemed to be in order.

It should be noted that some radioactive material is still buried at sea. A few countries in Europe, but mainly the United Kingdom, have been dumping solidified low-level radioactive waste matter into the deep ocean since 1967. This practice is limited to those nations with high population densities and little land space and is subject to restrictions of the International Dumping Treaty.

WHERE IS IT NOW?

The third method for disposing of low-level waste—shallow land burial—is the preferred method today. Only six shallow land burial sites are presently operated by commercial, private industry in the United States. These are located in New York, South Carolina, Illinois, Kentucky, Nevada, and Washington. With the exception of the Washington site, which is owned by the United States, all these facilities are on state-owned land. And each is regulated by the state in which it is located—except the Illinois site, which is regulated by the U.S. Nuclear Regulatory Commission.

In addition to the six commercially operated shallow burial sites, five such facilities are run by the Department of Energy. The sites are located in Idaho, Washington, New Mexico, Tennessee, and South Carolina. The accompanying map gives the names and locations of both the commercial and the Department of Energy disposal sites.

Given this kind of geographical distribution, it goes without saying that geology, water distribution, and climatic conditions differ markedly from one site to another. This is important, because these factors bear directly on the suitability of the facility for storing nuclear wastes. Rainfall, for example, which can contribute signifi-

MAJOR STORAGE AND DISPOSAL SITES FOR SOLID, LOW-LEVEL RADIOACTIVE WASTE

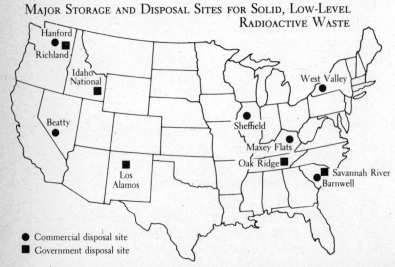

● Commercial disposal site
■ Government disposal site

cantly to any leakage problems that develop, ranges in these geographical areas from less than 3 to more than 46 inches per year.

The six commercial burial sites experienced no particular problems until the early 1970s. Thereafter, problems relating directly to water distribution occurred in two of them. In Maxey Flats, Kentucky, and West Valley, New York, water seeped into the trenches containing low-level waste material.

The average commercial burial trench measures about 30 feet in length and is 40 feet wide and 25 feet deep. Its volume is approximately 340,000 cubic feet, but not all the space is used because the containers of waste material do not touch. Only about half the volume of the trench is actually used, and after it is "filled," the excavation is covered with mounds of earth to prevent rain or snow from seeping through to the buried radioactive contents.

However, due to geological and meteorological features of the wetter eastern United States, precipitation can cause problems that would not arise in more arid regions. In Maxey Flats and West Valley, the earth used easily permits water to seep through it, and it appears almost impossible to keep water from getting into the trenches at these two sites.

As a result, the Kentucky site contributed radioactivity to the local environment during the early 1970s, and increased levels of tritium, cobalt-60, strontium-89, and strontium-90 were recorded. A program designed to manage water flow and distribution at this site was initiated for the express purpose of restricting radioactivity to the covered trenches.

At the New York facility in March of 1975, water seepage into two trenches produced elevated tritium levels in water samples taken at the actual site. The seepage was channeled into a reservoir, processed as low-level waste, and released. According to authorities, no significant increases in radioactivity were detected in the general environment.

Just how much of a hazard shallow burial of low-level waste represents to the general population is a matter of conjecture. What is certain is that these two sites did not meet the expectations of officials responsible for their existence and operation. The sites were authorized to operate in the belief that their characteristics and physical features were such that radioactive wastes would stay where they had been placed for hundreds of years. But in about ten years, radioactivity had already been detected off the site.

A WORD ABOUT RISKS AND RESPONSIBILITIES

Philosophically, it is important to remember that we do not live in a risk-free society. But a minimal risk is preferable to a greater risk. In disposing of radioactive waste with minimal risk of long-range effects, two factors must be considered. The first is the possibility that radioactivity may escape into the environment from where it has been stored—may leach out or be washed out into ground water, for example, and subsequently be ingested in food or drink. The second factor to consider is the radiation dose that would be delivered to humans in the event of such an escape.

According to expert opinion, even if high-level wastes leached into ground water, the slow rate at which this occurred would expose the populace to radiation doses not much greater than those we receive from natural sources. Accordingly, it is not necessary to strive for absolute assurance against escape. A more rational approach to nuclear waste disposal is to base decisions on the best available estimates of the probability of escape and of the consequences if escape does occur.

The fact of the matter is that most radioactive substances leached into surface water are transported to the oceans of the world without ever being ingested by humans. And after the material has flowed into the seas, the process of sedimentation and the remoteness of most ocean water from human activities combine to minimize the possibility of human ingestion.

Experts have concluded that, with state-of-the-art technology, it is now possible to dispose of nuclear garbage with only a negligible threat to the health of future generations. But all options must be kept open, so that we or our descendants can change to better methods of waste management should they become available.

We must also be careful not to saddle future generations unnecessarily with the burden of taking care of waste materials that we have generated. It is therefore vital that individuals and organizations charged with managing nuclear waste material demonstrate the highest levels of technical expertise and social responsibility.

6

Mutation and Cancer: The Major Effects of Ionizing Radiation

In the southwest corner of Utah, against a backdrop of rolling hills and mountains, lies the pretty little town of St. George. More than 15,000 people call it home, but the locale also draws tourists, many of whom spend their vacations in and around the town. Tourism, in fact, is the main business of St. George —except that fewer and fewer people are coming.

The trouble actually started during the twelve-year period from 1951 to 1963, when atomic weapons were being tested in the open air just to the west, across the Nevada state line. Nobody, including the various government agencies involved with radiologic health, could even guess how many rems of radiation wafted through the town's quiet streets during those twelve years. But according to a medical study, results of which were released in February 1979, the radiation levels were too high. Confirming the suspicions of many area residents, the study revealed that, during the test period, 2.5 times as many children in southern Utah died of leukemia as succumbed to the disease elsewhere, under normal circumstances.

Aside from putting a damper on tourism, the publicity catapulted St. George onto the national scene. News media descended on the town, and as TV mobile units and reporters prowled the streets, congressional hearings were held in Salt Lake City, with Senator Edward M. Kennedy presiding, and in Las Vegas, where the then Secretary of Health, Education and Welfare, Joseph A. Califano, Jr., promised that intensive studies would be undertaken on health problems in the area.

But the promises came too late for too many residents of St. George. Almost everyone who lived in the area during the test period has seen a member of his or her family die of cancer or fall victim

103

to cataracts, thyroid disease, various types of enlarged nodes, and any of a variety of disorders that may be associated with radiation exposure. Nor would St. George seem to be an isolated phenomenon. The town could very well have counterparts all through the West, where segments of the population were exposed to radiation as a result of either atomic testing or uranium mining and milling.

Nevertheless, it is difficult to establish a firm cause-and-effect relationship. Radiation produces no unique biological effects. To a physician, the burn an individual receives from high levels of radiation looks no different from the burn produced by, say, an open flame. Leukemia is leukemia, whether it is caused by chemicals, viruses, or radioactivity. Because of this difficulty in pinpointing an exact cause, scientists and physicians are unable to state with certainty that radiation is the culprit in any particular case of cancer.

What we do know is that ionizing radiation is *capable* of producing biological damage. Ionizing radiation has enough energy to jar an electron loose when it strikes an atom. The atom then becomes an ion, an entity that is chemically reactive and can therefore damage tissue. The types of ionizing radiation in the electromagnetic category are the x-rays and gamma rays produced by x-ray machines and radioactive decay, respectively. Ionizing radiations in the particulate category include the beta particles (negatively charged electrons), the alpha particles (helium nuclei with a double positive charge), and neutrons (uncharged particles associated with reactors, nuclear accelerators, and nuclear weapons.

Also remember that these types of ionizing radiation are not equally damaging. A specific amount of energy absorbed by tissue (measured in rads) can produce more or less biological damage (measured in rems), depending on the type of radiation. The difference, of course, is the intensity or "heaviness" of the radiation. It is analogous to the difference between being hit by a ping-pong ball and being hit by a bowling ball. It is called the linear energy transfer (LET). Dense particles, such as the alpha variety, constitute high-LET radiation, so they do more biological damage than such low-LET radiation as beta particles, gamma rays, and x-rays.

An x-ray exposure that makes your body absorb 1 rad of energy produces 1 rem of tissue absorption. (For this reason, when we are referring to x-ray dosage, the rad and the rem are essentially the same.) But a 1-rad dose of high-LET alpha radiation may be equivalent to a dose of 5 rems. Since the rem is a rather large quantity, scientists usually measure the extent of biological damage in millirems; a millirem is an amount equivalent to one-thousandth of a rem. With this in mind, it is possible to describe the actual biological effects of ionizing radiation.

The first case of human injury due to radiation that was reported in medical literature goes all the way back to 1895, the same year, quite logically, that Wilhelm Roentgen discovered x-rays. Only seven years later, physicians reported the first case of cancer caused by x-rays.

By the 1920s and 30s, a sizable body of evidence had accumulated to demonstrate the harmful effects of exposure to large amounts of ionizing radiation. This evidence was based on the experience of miners who were exposed to the radioactivity that filtered through their underground environments; individuals in special occupational groups who worked with radium; and early radiologists, including one martyr who went bald from permitting too many students to see "into his head."

This was an acute short-term effect resulting from large exposures. Ionizing radiation also causes more subtle, less noticeable effects that may not appear until years after the individual has been exposed. But the biological importance on a long-term basis of smaller, repeated radiation doses was not completely accepted until quite recently. It wasn't until Hiroshima that science really began to accumulate information about the biological effects of ionizing radiation.

IMMEDIATE EFFECTS

When a specific dose of ionizing radiation is administered quickly, it produces a more profound biological effect than the same dose given over an extended period of time. Area of exposure is also important. The more body area exposed, the greater the damage. Thus a dental x-ray has nowhere near the biological effect of a chest x-ray.

But basically, the damage done by ionizing radiation hinges on how much it affects the molecules that body cells need to function normally. Generally speaking, different types of cells show wide variation in their susceptibility to radiation damage. Cells of course, reproduce

Depending on its intensity, radiation slamming into a body cell may produce any one of four biological effects. The radiation may simply pass through the cell without causing any damage, like a rabbit zipping through a forest. Somewhat stronger radiation may damage the cell slightly, and the injured cell can repair itself. (Cells have the ability to correct some of the biological havoc wreaked by radiation, which is why a radiation dose administered in fractions over extended periods of time causes less damage than the same dose given in one fell swoop.) Higher radiation intensity may cause more severe damage, so that the cell not only fails in its efforts at self-repair but also reproduces in a damaged form for long periods of time, producing unhealthy progeny. Finally, heavy radiation can simply kill a cell, the way a bulldozer going through a forest will certainly kill some trees.

by dividing, a process called mitosis. As a rule, those cells that divide rapidly are more sensitive to radiation effects. And cells that are primitive or nonspecialized are more sensitive than highly specialized cells. White blood cells, or lymphocytes, are highly susceptible— probably the most susceptible to radiation of all the body cells. Other varieties of white blood cells are somewhat less sensitive, followed by young red blood cells. Certain body tissue, such as that which covers an organ, is moderately sensitive to radioactivity, whereas the cells that make up muscle and nerve tissue exhibit little sensitivity to radiation exposure.

STAGES OF RADIATION EXPOSURE

Such exposure usually follows a specific course that can be broken down into three stages. The first of these is the latent period, which occurs between the actual exposure to radiation and the onset of the first detectable effect. The latent period may vary considerably in length. If effects occur within a matter of minutes, hours, days, or even a few weeks, physicians refer to them as short-term effects. Long-term effects, on the other hand, do not show up for years or even generations.

After the latent period (which may last anywhere from minutes to decades), the effects on cells or tissue actually take place. These effects (we will discuss them in detail in a moment) usually come about because living tissue exposed to radiation stops reproducing.

Depending on how much and what type of radiation has been absorbed, the cessation of reproduction may be temporary or permanent. Other disruptions of normal cell function also occur. Many of these effects are not related solely to radioactivity. Individually, they can be produced by other means—chemical substances, for example. But it is the *combination* of effects that is unique to radiation. No known chemical substance causes exactly the same constellation of effects.

The third stage, called the recovery period, is based on the tissue's ability to heal itself under certain conditions. If effects are short-term ones, recovery usually takes place to a greater degree. But some irreparable damage may occur that can later be responsible for long-term effects.

ACUTE RADIATION SICKNESS

When a dose of radiation is delivered to the body over a very short time span, the biological effects are short term and are called *acute somatic effects.* These affect the health of only the person who was exposed; they are obviously not communicable.

Much of our understanding about the somatic effects of ionizing radiation dates from the climactic days of World War II and the dropping of atomic bombs on Hiroshima and Nagasaki. Individuals who survived the blast absorbed extremely high radiation doses in very short periods of time. Many received doses as high as 50 rems —100 times today's U.S. allowable annual dose—in less than an hour. This set of circumstances almost guarantees death or a condition known as acute radiation sickness within a few days or weeks after exposure. Victims of the first atomic bombs who absorbed less radiation and could be observed medically for the rest of their lives demonstrated a significantly greater incidence of leukemia, cataracts, and cancer than might normally be anticipated in a similar group of people who had not been exposed. Because these effects afflicted the individual directly exposed, they are considered somatic (though late) effects. We will have more to say about this later.

When an individual absorbs the dosages that were common at Hiroshima, the first signs of acute radiation sickness appear during the prodrome, or initial phase of the illness. They consist of nausea, vomiting, and a general sense of not feeling well. Following the prodrome comes a latent stage during which the victim may feel better. However, changes taking place inside the organs concerned with blood production and in other parts of the body will bring on the next stage of the syndrome.

In this third phase, the full-blown disease emerges. Among the symptoms associated with high radiation exposure are fever, lowered resistance to infection, hemorrhage, loss of hair, severe diarrhea, disorientation, and disturbances in function of the cardiovascular system. Which of these symptoms appear and to what degree of intensity they occur usually depend on just how much radiation the individual received.

In the fourth phase of acute radiation sickness, a person either recovers or dies.

The type and severity of acute radiation sickness are directly related to dose and to the sensitivities of various types of exposed cells. Thus, at relatively low doses, biological injury is most likely to occur in cells with highest sensitivity, such as white blood cells. When these cells are injured, the usual effects are fever, infection, and hemorrhage.

Higher doses also damage cells that are less sensitive to radiation. Such cells line the gastrointestinal tract. When they are destroyed, fluid is lost from the digestive system, resulting in bacterial infection and severe diarrhea.

When dosage is extremely high, the more resistant cells, usually found in nerve tissue, are injured. With the central nervous system no longer functioning normally, an individual may become severely ill very suddenly, with disorientation and shock.

Though individuals vary tremendously in their response to radiation, no overt signs of illness are generally detected at doses of radiation under 50 rads. Doses in the 100-rad range are not usually associated with symptoms, but subtle changes in the blood can be detected. At 200 rads, the first definite signs of injury appear (fever, hemorrhage, and hair loss), and extremely sensitive individuals may even die after such exposures.

A dose of 450 rads is considered the median lethal dose. That is, half the people exposed to this dose of radioactivity die, usually as a result of infection. Doses of 600 rads produce gastrointestinal symptoms and are fatal for most people. Doses from 800 to 1,000 rads are always fatal. At higher doses, the central nervous system exhibits symptoms, and death results from cardiovascular collapse.

DEVELOPMENTAL EFFECTS

In addition to acute somatic effects, exposure to radiation can produce what are referred to as *developmental effects.* These abnormalities befall fetuses in utero or young children. During early fetal development, radiation doses as low as 5 rems can cause bone malformation and nervous system disorders in children born to exposed women.

In Hiroshima and Nagasaki, pregnant women who survived the explosions bore children with an abnormally high incidence of birth defects. These youngsters also demonstrated a much greater tendency to develop cancer or leukemia as they grew up. In addition, these pregnant women miscarried or produced stillbirths at a significantly higher rate than average.

As for children exposed to radiation in the first atomic holocausts, they grew up apparently much more susceptible to health problems than adults who had received comparable doses. And regardless of the age of the survivor, exposure to radiation resulted in an increased tendency to develop cancer or leukemia, and to produce deformed children.

Congenital defects have been studied extensively in laboratory animals receiving varying doses of radiation, and it is clear that exposure of a fetus or embryo to radiation may result in death, birth defects, or effects that are not evident at birth but develop years later.

Whether exposure causes death of the fetus or specific major abnormalities in development depends to a great extent on when during gestation the exposure occurs. Immediately after conception, the embryo consists of just one cell—the fertilized ovum—which will divide into other cells. Damage to any of the cells in the embryonic stage eventually causes more havoc than damage to the cell of an adult, who is composed of some 50 trillion cells.

In 1968, a study was published on the subject of abnormalities in children exposed to ionizing radiation during various stages in the 9-month gestation period. The investigators translated their findings into a kind of timetable of radiation injury to the fetus. For example, radiation doses of more than 250 rads before two to three weeks of gestation have elapsed cause the death of a majority of fetuses. For some inexplicable reason, however, surprisingly few fetal abnormalities are seen if the fetus survives this dose.

After the first month of gestation, but before the end of the third month, a radiation dose received by the fetus can injure a large number of structures, especially bones and tissue of the central nervous system. During the fourth month, irradiation of the fetus usually results in mental retardation and a congenital birth defect known as microcephaly—abnormally small size of the head and brain. After five months of gestation, the fetus has developed to such an extent that it is more resistant to abnormalities and death, but radiation doses during this stage can still result in birth defects.

A study done in 1930 underscored the extreme vulnerability of the fetus to radioactivity during the first trimester of pregnancy. In this study, the children of 106 women who had undergone various forms of therapeutic irradiation during the first trimester of pregnancy were painstakingly observed. The investigators noted that slightly more than a third of these children (28 out of 75) were malformed. Included were bone defects and nervous system defects that produced such conditions as microcephalic idiocy, mental retardation both with and without physical deformity, water on the brain (hydrocephaly), Down's syndrome, defects in the spinal column, clubfeet, limb malformation, blindness, and other eye abnormalities.

Some fifteen years later, similar studies were carried out to specifically examine children born to women who were pregnant and survived the atomic bomb blast at Hiroshima. Among 11 women who were known to have absorbed large radiation doses, 7 offspring suffered from microcephaly and mental retardation. Children of mothers who were further away from where the bomb exploded did not appear to be predisposed to develop microcephaly. At Nagasaki, the fate of 30 fetuses who had received radiation doses as a result of the bomb blast was traced. The investigators noted 7 fetal deaths, 6 deaths at birth or shortly thereafter, and 4 mentally retarded offspring.

These findings assume contemporary significance when a woman who does not realize she is pregnant undergoes irradiation in a diagnostic radiologic examination. Possible pregnancy should always be taken into account in women of childbearing age when diagnostic x-rays are considered, especially x-rays of the pelvis, abdomen, or lower back. To avoid irradiation of a woman who doesn't suspect that she is pregnant, most physicians resort to what is known as the 10-day rule. This guideline limits x-ray procedures on women of childbearing age to certain specific days during the menstrual cycle, to preclude the possibility of pregnancy. Women whose physicians do not take this precaution should mention their desire to observe the 10-day rule when any x-ray procedures are scheduled. According to this rule, x-rays should be taken only during the first 10 days of the menstrual cycle (day number one is when menstrual flow begins).

Of course, procedures that expose a woman to radiation are sometimes necessary, even though she is pregnant. This must be done with extreme care; it is generally believed that no dose level is completely safe. What complicates our effects to assess the risks is that congenital abnormalities occur in about 5 percent of the population under normal circumstances. Hence it may not follow that a given birth defect can absolutely be attributed to a small dose of radiation. All physicians can say with certainty is that radiation seems to increase the *probability* of a birth defect and that this increase varies with dosage.

The cutoff figure used most often in scientific circles is 10 rads. Thus many authorities recommend that, when a fetus has been exposed to as much as 10 rads of radiation during the first month and a half after conception, a therapeutic abortion should be performed due to the high probability that a defective child will otherwise be born. This may be an extremely difficult decision to make for a variety of reasons. The relative merits of the procedure should be thoroughly discussed on a professional level and counsel sought from whomever the couple feels can contribute to an acceptable solution.

LATE EFFECTS

Actually, the acute high doses that produce immediate effects are not seen often. Of more practical interest are the biological effects of ionizing radiation that arise years later in individuals who have survived high doses. Because years may elapse before damage becomes apparent, these effects are called the late effects of radiation, and they can be produced by low doses of radiation administered over extended periods of time—as when patients undergo periodic diagnostic x-ray procedures or radiation workers are exposed on the job.

GENETIC EFFECTS

Some of the most insidious of these late effects are genetic effects. Whereas somatic effects directly influence the health of the people exposed, genetic effects are their legacy to their heirs. No entirely convincing data exist on how ionizing radiation actually damages human genes, but we may be able to put this discussion in some kind of perspective by talking a little about cell biology.

Each human organism begins life as a single cell resulting from the fusion of an ovum from the mother and sperm from the father. All traits inherited from both parents are transmitted by this fertilized ovum. The "software," or developmental instructions of the ovum are contained in its nucleus on ribbonlike structures known as chromosomes. As the embryo develops by cell division, all the genetic information in the original, fertilized ovum is faithfully duplicated. Hereditary information is passed from generation to generation via a genetic code, which is embodied in the structure of what are called *genes*.

The chromosomes in the nuclei of cells are made up of genes, more than 1,200 strung beadlike in every chromosomal ribbon. Genes have been called the "units of heredity" and are usually referred to by their effects on the body. We speak, for example, of a gene for red hair. The number of genes in a human being has been estimated to be approximately 20,000.

Genes owe their activity to a substance called deoxyribonucleic acid, or simply DNA. All genetic information is apparently contained in the very long and physically complex structure of the DNA molecule. Altering the DNA has been compared to shooting a bullet into a computer: it is virtually impossible to predict what effects will result. Disturbance of the data carried by DNA is called a *mutation,* and anything that causes a mutation is called a *mutagen.* Among various mutagens are chemicals and—you guessed it—radiation.

Chemicals can produce changes in the hereditary apparatus of a cell, but damage is usually limited to cells with which direct contact is made—those in the skin, digestive tract, liver, and so on. Chemicals rarely come in contact with the gonads, or organs of reproduction, where sex cells are produced. Thus the action of chemicals on tissue primarily threatens the individual alone, not her or his off-spring.

Radiation is another matter. By its nature, it is pervasive and nondiscriminatory. It can affect any part of the body. Because of the deep-penetrating ability of x-rays and gamma rays, the inside of the body is often just as susceptible to radiation exposure as the outside. And this susceptibility to radiation includes the gonads. Like chemicals acting directly on tissue, ionizing radiation can produce mutations in living cells that make them cancerous. But it can also cause mutations in sex cells. When this occurs, unborn generations are threatened.

A mutation in a sex cell is a basic alteration in heredity, and it produces offspring essentially unlike their parents. Often, sex cells are produced with an extra chromosome enclosed (47 instead of the 46 that normally occur in humans). Today we know that, when this mutation produces an extra chromosome at a specific site in the cell, a condition known as Down's syndrome (or as it is commonly but inaccurately called, mongoloid idiocy) results in the offspring. Most mutations, however, are not the result of changes in chromosomal material. Rather, they result from less spectacular alterations in the chemistry of genes composing the chromosomes. And most mutations produce fairly subtle effects, such as relatively insignificant metabolic changes that may affect how certain foods are used by the body, an increased tendency to develop a certain disease, or perhaps just slightly lower intelligence. Furthermore, many mutations are *recessive*. That is, they do not appear in an individual unless *both* his or her parents passed along the same mutant gene.

MUTATION AND EXPOSURE TO RADIATION

The ability of ionizing radiation to produce mutations was discovered in experiments on the fruit fly in 1927. These and other laboratory studies over the years constitute our most reliable source of data on the genetic effects of radiation. (The largest group of humans on whom genetic studies can be carried out are the descendants of those who survived the Hiroshima and Nagasaki A-bomb blasts. To date, no increase has been noted in infant mortality or incidence of malformation.)

From the experimental work on insects and animals have come two widely (but not universally) accepted generalizations about radiation and human health. First, even the smallest radiation dose will produce some mutations in chromosomes of gonadal cells that become sex cells—there is *no* dose below which genetic damage fails to occur. And second, the exact extent of mutational damage depends on the radiation dose and the length of time over which it is absorbed.

If our civilization is to remain healthy and thrive, the fewer mutations the better; mutations essentially represent disorders. But the tendency of today's technology is to *increase* the mutation load through increased exposure to mutagens and through superior medical care, which saves and prolongs the lives of people with genetic problems. If these individuals reproduce, their genetic defects are exhibited in more and more people, and the ultimate effect is a weakening of the human race. These facts make it extremely important that new mutations be kept to a minimum. And when you consider the unusual ability of ionizing radiation to cause mutations, a vital scientific objective should be the elimination of unnecessary irradiation to the human organs of reproduction.

IONIZING RADIATION AND CANCER

In addition to those effects that involve heredity, another late consequence of ionizing radiation is the somatic effect that results in an increased incidence of cancer. Here the effects of the radiation exposure are visited directly on the person irradiated, rather than on some future descendant. The problem that faces scientists, however, is how to establish a direct link between radiation exposure and cancer.

This problem is complicated by several factors. For example, cancers caused by exposure to radiation are no different from those that occur under natural circumstances. If a person who absorbs a high level of radiation develops cancer thirty years later, the disease may well be a result of the exposure, but the relationship cannot be proved. This is because so many other variables enter into the picture: the type of cancer, its site in the body, the sex and age of the victim, nutritional status, and occupational history, to name a few.

Cancer research on humans is also made more difficult by the low incidence with which cancer of any one type strikes. Too, there are

CANCERS LINKED TO RADIATION IN PARTICULAR POPULATIONS +

TYPE OF CANCER	ATOM BOMB RADIATION			MEDICAL RADIATION									OCCUPATIONAL RADIATION			
	Japanese atom bomb survivors	Marshall Islanders	Nuclear test participants	Ankylosing spondylitis (x-ray)	Ankylosing spondylitis (radium)	Benign pelvic disease	Benign breast disease	Multiple chest fluoroscopy	Enlarged thymus (infants)	Thyroid cancer (I-131)	In utero x-ray	Diagnostic x-ray	Radium dial painters	Radiologists	Uranium & other miners	Nuclear workers
Leukemia	***		*	***		*			**	*	***			***		*
Thyroid	***	**							***							*
Female breast	***						***	***								
Lung	***			***											***	
Bone	**				**								***			
Stomach	**			**												
Esophagus	**			**												
Bladder				**												
Lymphoma (including multiple myeloma)	**			**										**		*
Brain										*				**		
Uterus						*										
Cervix	*															

TYPE OF CANCER	ATOM BOMB RADIATION			MEDICAL RADIATION									OCCUPATIONAL RADIATION			
	Japanese atom bomb survivors	Marshall Islanders	Nuclear test participants	Ankylosing spondylitis (x-ray)	Ankylosing spondylitis (radium)	Benign pelvic disease	Benign breast disease	Multiple chest fluoroscopy	Enlarged thymus (infants)	Thyroid cancer (I-131)	In utero x-ray	Diagnostic x-ray	Radium dial painters	Radiologists	Uranium & other miners	Nuclear workers
Liver	*															
Skin	*													***	**	*
Salivary gland				*	*				**							
Kidney									**							
Pancreas	*												*			
Colon						**										
Small intestine						*										
Rectum						**										

+ Strong associations are indicated by ***, meaningful but less striking associations by **, and suggestive but unconfirmed associations by *

*Adapted from *Interagency Task Force on the Health Effects of Ionizing Radiation: Report of the Work Group on Science*, U.S. Department of Health, Education and Welfare, June 1979.

few groups of irradiated individuals large enough to provide statistically significant data on tumor types or sites. Even when such groups of people exist (Japanese A-bomb survivors, for example), the long years that characterize the cancer latency phase defeat the possibility of following up on these populations to see if and when tumors develop, and they complicate the task of keeping an accurate history of any one patient's exposure to radiation. An increase in the incidence of cancer traceable directly to radiation exposure may be missed unless a particular study is carried on for decades. Further, much of the data that have accumulated on radiation-induced tumors in humans come from individuals exposed to internal radiation (by eating certain foods, for example) that may vary considerably from time to time and place to place.

RADIATION AS A CARCINOGEN

Specially designed experiments using animals and the proper radiation doses have demonstrated that ionizing radiation can have a cancer-causing (carcinogenic) effect. Even so, the exact mechanism by which radiation causes cancer is hard to pin down. Like other cancer-producing agents, radiation may be only one of many factors that must be present in a particular organism before cancer results.

One theory suggested by these animal studies is the virus theory, which attributes cancer to a specific virus that, when irradiated, attacks normal cells and gains access to their nuclei. Once inside, the virus causes the cell to reproduce prodigiously, substituting the genetic information of the virus for that of the cell and thus producing a cancerous tumor. This theory would also work if it were the cell's resistance to the virus that radiation destroyed, permitting the virus to enter the cell with the same net result.

Another theory involves chromosomal damage. Diseases like leukemia are known to be associated with injury to the chromosomes. It is therefore possible that radiation exposure is the cause of chromosomal damage, which in turn kicks off the disease process.

Because ionizing radiation is a powerful mutagenic agent, another theory suggests that cancer is produced directly by the mutation of a cell in normal tissue. As neat as this theory sounds, it is probably too simplistic. Cancer resulting from a single mutation would seem unlikely. A series of events is probably involved in the development of cancer, and a cell mutation may contribute to the series as one link in a chain.

Evidence of the cancer-causing effects of radiation is not limited to laboratory animals. The Hiroshima and Nagasaki survivors, who were exposed to radiation doses in the 100-rem range, actually did contract leukemia at a significantly higher rate than might normally be expected. This higher incidence was predictable on the basis of the higher radiation dose, thus making a strong case for the leukemia-causing potential of radiation. Breast cancer and cancer of the thyroid gland also appeared more frequently than normal among the atomic bomb survivors.

Other evidence of the link between radiation and cancer in humans also exists. Studies done on uranium miners in the early 1900s indicate that they developed lung cancer at a 50 percent higher rate than the general population due to inhaling radioactive dust.

Around the same time, there were reports of dentists and x-ray technicians dying as a result of skin cancer. Unaware of any risk, these individuals did not think of protecting themselves from the streams of x-rays passing through their bodies. Of particular interest was the number of dentists with cancerous lesions of the fingers ordinarily used to hold dental films in place in their patients' mouths.

Early in this century, one of the few employment opportunities for young women was painting luminous numerals on watches and clocks. These women, who inadvertently ingested radium paint while performing their delicate jobs, became more likely to develop bone cancer and other malignancies than their counterparts in other lines of work.

Cancer was also found to occur more often in patients who were treated with large x-ray doses to relieve the pain of a disabling arthritic disease of the spine; in children treated with x-rays for enlarged thymus glands, swollen tonsils and adenoids, and acne; in patients treated with x-rays for ringworm of the scalp; in children whose mothers underwent pelvic x-ray examination during pregnancy; and in patients who were treated for tuberculosis by procedures done with fluoroscopic assistance.

In determining the risk of cancer, estimates must be based on data available from studies done on large human populations, such as the Japanese atomic bomb survivors. Though this information is not yet complete, we know that, for these victims, the number of all forms of cancer, above what might ordinarily be expected, corresponds to 50 to 78 deaths per million people exposed, per rem, during the 20 years between 1950 and 1970.

A risk estimate can be thought of as the difference between cancer risk in an irradiated population, like the A-bomb survivors, and the corresponding risk in a similar group of people who were exposed only to normal levels of radiation. There is no simple way for *individuals* to calculate their chances of developing cancer following exposure to large radiation doses, such as the Three Mile Island incident, or low doses, such as too many x-rays. Biostatisticians concern themselves instead with *populations* and total exposures to these populations.

Hence all we can offer is a qualitative rather than a quantitative discussion of radiation-caused cancer. You must not try to extrapolate personal risk when so many factors enter into the development of such cancer. Remember that long latency periods make it difficult to establish a correlation between exposure and appearance of the disease. Also remember that the natural incidence of cancer varies markedly from organ to organ and is influenced by genetics, age, sex, geography, diet, socioeconomic background, and many other factors. Finally, the amount of radiation that will produce cancer in one person (termed the *individual dose threshold*) depends not only on that dose but also on the extent to which nonradiation factors contribute to the total carcinogenic process.

CANCERS ASSOCIATED WITH EXPOSURE TO RADIATION

As we have noted, data obtained from studies on Hiroshima and Nagasaki survivors and on patients treated with high doses of radiation indicate an increased incidence of most forms of leukemia. And the data suggest that the possibility of developing leukemia increases in those who received the radiation as fetuses or young children.

The incidence of thyroid cancer has also been found to increase with increasing doses of radiation. The thyroid gland is sensitive to radiation, but susceptibility is higher in females than in males and higher in children than in adults. Thyroid cancer has been seen to develop in children following very low radiation doses. Most thyroid tumors induced by radiation usually appear within about twenty-five years. However, thyroid cancer is not a major killer; only about 3 or 4 percent of cases are fatal.

Lung cancer is known to be an occupational disability for uranium miners who inhale radon gas. (Cigarette smoking, air pollution, and chemicals such as asbestos are also cancer-causing agents.) An unquestionably higher incidence of lung cancer has been noted in the uranium miners of Colorado and Czechoslovakia, but it usually takes between fifteen and twenty years of exposure before the first symptoms appear. Although most of it is caused by the victim's inhaling radionuclides that emit alpha particles, evidence also exists of a higher lung cancer rate in people exposed to external sources of x-rays or gamma rays. Death from lung cancer related to radiation befalls about 1 out of every million people exposed to 1 rem per year.

Bone cancer is rare; mortality related to radiation is approximately 1 death for every 5 million people exposed to 1 rem per year. This disease usually requires that the victim ingest radioactive isotopes (such as radium), as painters of luminous watch dials might do if they were to lick their brushes to sharpen the points. An important factor in the development of bone cancer is the age of the individual at the time of exposure. The younger the patient, the higher the risk.

Radiation-related skin cancer was first seen in early radiologists who did not bother to shield their hands when doing fluoroscopic examinations. Here the onset of cancer tumors followed years and years of a chronic skin rash. The susceptibility of skin to radiation is considerably lower than that of such other tissues as thyroid and blood-producing bone marrow. Oncologists, physicians who specialize in cancer, believe that skin cancer will not result unless the individual receives an acute ionizing radiation dose in excess of 1,000 rems.

For cancer of the digestive system, such as stomach cancer, death attributable to radiation will probably strike 1 out of every million people exposed to 1 rem over a year's time.

We now come to breast cancer, a topic that is widely discussed today. Unlike other types of cancer, breast cancer is very closely associated with radiation. After exposure—usually in the form of periodic mammograms (diagnostic x-rays of the breast tissue)—cases of breast cancer are seen within ten years and at a higher frequency than might normally be expected. New cases may continue to develop for up to thirty years. The average latency period is probably about twenty-five years.

Records indicate that many women have received frequent mam-

mographic examinations, often as many as forty or fifty, each exposing them to doses of 4 to 20 rems per examination. As might be
expected, such women show a higher propensity for developing breast
cancer. Repeated low-dose radiation, such as might be absorbed during routine chest x-rays or low-dose mammograms*, actually *increases*
the woman's chances of developing breast cancer. The doses are
probably cumulative, so many small doses may subject the patient to
as much risk as one large dose. Of course, one's risk chance of
developing breast cancer may be increased by other factors, such as
a history of past inflammation of the breasts, never having delivered
a child, and having received one's first x-ray exposure to the breasts
at an early age.

The other side of the coin is that, through x-ray mammography,
breast cancers can be detected before any symptoms are noticed. And
if breast cancer *is* detected early, it is more likely that the patient will
survive. This is why many authorities advocate x-ray mammography
for mass screening of women who have no symptoms. Obviously, it
is necessary to strike some kind of balance between the benefits of
early detection of breast cancer and the risk of actually causing it to
develop in later life.

Many women find themselves on the horns of this dilemma, and
there are no easy answers. Right now, the pendulum seems to be
swinging in the direction of mammography. In an article that appeared in the *Journal of the American Medical Association* in late
1979, the author stated his opinion that, while radiation screening may
induce some breast cancer, the risk is extremely small and seems worth
taking, in view of the number of cancers that can be detected early.

According to this report, one woman a year out of a million women
exposed to radiation from low-dose mammography develops breast
cancer ten years after the exam. But the *natural* incidence of breast
cancer is much higher—ranging from about 1,000 cases a year per
million forty-year-old women to approximately 2,000 cases a year per
million women in their seventies. Considering that half of these cases
might be detected by mammography at an early curable stage, the
risk from mammography seems small—about 1 case induced for
every 500 detected in forty-year-old women.

*Low-dose mammography is a relatively new x-ray technology that allows the total
radiation dose to be reduced.

Your doctor is the most reliable source of advice on this subject. He or she knows you, your family history, and your predisposition to other diseases and is also aware of the latest information on this issue.

In the absence of firm evidence establishing a measurable correlation between tissue damage and low-level radiation dosage, the wisest course of action is to assume that the higher the dose, the greater the damage. Many scientists believe there is no threshold level of radiation exposure below which danger ceases to exist and that *any* radiation dose, regardless of size, subjects the person exposed or any potential offspring to a certain amount of risk.

This belief is termed the *linear nonthreshold dose-response hypothesis.* Although this hypothesis probably overestimates the amount of x-ray–induced biological damage caused by ionizing radiation (especially at very low doses), it would seem to be a good touchstone when trying to evaluate your personal risk. Always assume that biological damage can originate in a single cell and that every dose of radiation absorbed, no matter how small, *could* play some part in inducing cancer, even if that cancer appears twenty-five to thirty years after exposure.

7
Even Nonionizing Radiation Can Be Hazardous

If ionizing radiation alone was hazardous to human health, those of us fortunate enough not to become victims of nuclear attack or accident and cautious enough not to work with or live near sources of radioactivity might put the subject out of our minds. But there is more to think about. Also possibly dangerous is nonionizing radiation—the kind you may be exposed to in your own home from the likes of microwave ovens, heat lamps, garage door openers, tanning lamps, and many other devices you use every day.

As we discussed in Chapter 2, nonionizing radiation does not have enough energy to jar electrons loose but simply produces an energy shift. This causes excitation of the atoms and molecules, makes them chemically reactive, and usually generates heat. The types of electromagnetic radiations that do not cause ionization are less energetic than x-rays—particularly the radar, microwave, infrared, and ultraviolet wavelengths—but all can cause problems.

Nonionizing radiation is converted into heat when absorbed by tissue, and though the temperature elevation does not cause electrons to break loose and ionization to occur, it can still produce biological damage. In addition to the thermal effect of microwave radiation on tissue, evidence from laboratory research with insects and small animals shows that microwave energy can sometimes induce changes in the structure of certain chromosomes and thus produce effects similar to those of ionizing radiation.

Exactly how nonionizing radiation interacts with living tissue is still not completely understood. We do know that, when parts of the body receive microwave or radar radiation, heat is developed within the tissues and then carried by the bloodstream to the body surface, so that it can be dissipated. If microwave exposure is so great that the heat produced cannot be adequately reduced, tissue will be destroyed and death may result. The actual outcome depends on such factors as level of radiation, length of exposure, and specific area of the body exposed.

DANGERS ASSOCIATED WITH MICROWAVE AND RADAR RADIATION

Just how much microwave radiation can the body stand? Scientists are not sure. In 1947, Dr. Herman Schwan, a research scientist specializing in electrical engineering and physical medicine, embarked on a study to determine a safe level for humans. Working on the assumption that heat was the only effect caused by microwave radiation, and taking into consideration certain other biological processes involved with the body's ability to dissipate heat, Schwan concluded that a human being can safely withstand a microwave radiation dose of 10 milliwatts per square centimeter of body surface.

Schwan's conclusion was confirmed by experiments in which animals were exposed to higher and higher microwave doses until actual burns resulted. A level somewhere below that which produced burns was presumed to be safe. On the basis of this work, the 10-milliwatt standard was generally accepted as safe. This standard was adopted by the armed forces for its personnel involved with radar and by the American Standards Institute as an occupational guideline.

This figure had never actually been tested for any other mechanism but heat. Of particular interest is that the level *we* presume safe is a *thousand times greater* than the microwave standard in the Soviet Union, where the permissible level is only 0.01 milliwatts per square centimeter. The 10-milliwatt level is still used as an index of safety in the United States today, despite growing conviction in the scientific community that it is too high.

ANIMAL STUDIES

That the effects of nonionizing radiation in the microwave range partially result from the body's inability to dissipate heat has been demonstrated experimentally with laboratory animals. As different doses of microwave radiation at varying wavelengths are administered to laboratory animals for specific periods of time, various physiological effects occur. Redness of the skin, prostration, convulsions, swelling of the head and genitals, and death have been reported. As the level of microwave radiation and the length of exposure time are increased, adverse effects appear more rapidly and are more severe. Another interesting point is that microwave radiation in "pulsed," or noncontinuous; such doses appear to produce more harmful biological and physical effects than radiation that pours out in a steady stream.

Results from animal experiments are not necessarily applicable to humans. Differences in physiology, physical dimension, and form make it next to impossible to predict how closely the effects in animals will resemble those in humans. But certain qualitative conclusions can be drawn from the effects of microwave radiation on animals.

As we have seen, microwave intensity is measured in milliwatts per square centimeter. According to some reports, intermittent exposure of animals over long periods of time to low doses of below 1.5 to 30 milliwatts per square centimeter will produce a series of minimally harmful effects that eventually add up to serious damage. This type of cumulative exposure has been associated with behavioral and thought defects in laboratory animals.

Damage to the testicles is also seen in male animals exposed to microwave radiation levels high enough to cause temperature increases. In females, similar injury is produced in the ovaries, resulting in altered fertility and a higher incidence of stillbirths. Temperature effects also seem to influence development of the fetus, which is more susceptible to microwave radiation in the early stages of its growth. An elevation in temperature of 2.5 to 5 degrees centigrade above the normal temperature of the species, sustained for an hour, appears to be the level at which defects in growth and development occur.

As for genetic effects, the scientific literature contains numerous references to chromosome damage and abnormal cell reproduction brought about under certain conditions and in certain cell types by nonionizing microwave radiation. But the ability of such radiation to produce genetic effects in mammals has yet to be proved.

One of the major questions about microwave radiation is whether it can cause biological injury exclusive of the thermal injury known to occur at sufficiently high levels. An example is red blood cell defects. Here the nuclear structure and reproductive capability of cells may be directly affected. Definitive answers to these questions are needed for proper assessment of the leukemia or other cancer-causing effects of microwave radiation.

A combination of both heat and nonthermal mechanisms is probably responsible for one of the most common effects of microwave exposure—cataracts and damage to the lenses of the eyes. Such effects are presumed to result from repeated exposures.

EFFECTS IN HUMANS

The effects of nonionizing radiation in humans have not been studied extensively. The subjective effects most often reported include headache, insomnia, irritability, loss of appetite, and faulty memory. Vague feelings of heaviness in the head, drowsiness, and chest pain may also occur. Physiological effects include easily irritated skin, excessive sweating, fluctuating blood pressure, and changes in various eye tissues such as the retina. In the area of emotional effects, anxiety has been reported, as well as hypochondria and the harboring of suicidal thoughts. Slow heartbeat (a pulse rate of 60 per minute or lower) has also occurred.

As for sterility or infertility, exposure to microwave radiation would seem to be a significant factor. One case involved a man who had fathered children but became sterile while working with radar equipment that emitted nonionizing radiation in excess of 3,000 times the accepted safe levels and who did not wear protective gear. A precise causal relationship could not be firmly established, because no pre-exposure examination had been done. But one of the organs most susceptible to damage from radar and microwaves in human males is the testicle.

Another highly susceptible organ in humans, as in animals, is the eye. And microwave or radar radiation in sufficient doses over sufficient periods of time can result in the formation of cataracts. Parts of the neuroendocrine system that involve nerves of the various glands in the body are known to be sensitive to small changes in temperature. Microwave exposure is therefore capable of damaging these sensitive areas.

According to some researchers, higher microwave exposures have been associated with increased production of red blood cells and lymphocytes, cells formed in lymph tissue. When white blood cells are exposed to low levels of radiation, their ability to neutralize cells that are harmful to the body declines. This would reduce the body's ability to resist disease, but the effect is of short duration. (Conflicting evidence suggests that elevated temperature in humans is part of the body's response to infection through bactericidal action.)

Several interesting studies have been done on large groups of people believed to have been exposed to microwave or radar radiation. While no ill effects were noted in radar workers in the United States after a four-year study, similar groups in eastern Europe suffered altered nervous system and cardiovascular function.

A higher incidence of Down's syndrome, which results from a mutation affecting chromosomes, has been reported. Subsequent study, however, failed to establish a definitive link between Down's syndrome and exposure to radar waves.

In Poland, an extensive study lasting twelve years was done on more than 800 men between the ages of twenty and forty who were exposed to microwaves in the course of their work. The purpose of this study was to determine whether any defects or disorders had emerged that might preclude further occupational exposure. In two groups, one exposed to higher levels of radiation than the other, no greater incidence was seen of cancer or disorders of the nervous system, ocular lens, blood system, or endocrine system. Other functional disturbances were seen, but not in large enough numbers to be directly related to the occupational microwave exposure.

An interesting study done in the Soviet Union showed that workers exposed to both radar and microwave radiation seemed to develop certain white blood cell abnormalities, regardless of the degree of their exposure. Biochemical changes in other components of the blood, and an increased incidence of eye problems were also noted. The main symptoms, however, occurred in the central nervous system and the cardiovascular system: slow heartbeat, irregularity of heart rhythm, and fluctuating blood pressure.

A chronic overexposure syndrome described by Russian researchers is characterized by patient complaints of headache, irritability, insomnia, weakness and lassitude, loss of sexual libido, and undefined feelings of not being well. This syndrome seems to appear when overexposure levels are greater than 10 milliwatts per square centimeter. This is only one ten-thousandth of the power needed to light a

100-watt household bulb, spread over an area smaller than a postage stamp.

Apparently, when healthy adults are occupationally exposed to microwaves at levels no higher than 1 milliwatt per square centimeter, biological injury does not occur. However, acute exposures at levels considerably higher than 10 milliwatts per square centimeter *can* injure important organs and lead to nerve disability, perhaps even death. A report appearing in the medical literature *(Eye, Ear and Throat Monthly,* July 1975) described just such a case. The patient worked as a microwave troubleshooter from age twenty-one in 1942 through age forty-eight in 1970. After extensive exposure to high-power microwaves, he eventually became both blind and deaf.

An important point to keep in mind is that virtually all information available on the biological effects of nonionizing radiation has been accumulated through studies involving normal adults. Science is largely ignorant of how children are affected by intermittent or continuous exposure—especially children who live in the vicinity of radar installations or television transmitters. We do know that, because of the size of their bodies and different proportions, children absorb nonionizing radiation differently from adults. More study is urgently needed.

MICROWAVE RADIATION AND CANCER

One of the most vital areas of concern is any possible link between nonionizing radiation and cancer. Apparently such links do exist. Scientists believe that low-frequency, nonionizing radiation may be associated with disruption in the balance of body hormones. And such hormonal imbalances have been implicated in the development of cancers of the breast, cervix, and prostate. The growth of these cancers is thought to be speeded up or slowed down by hormones.

Another possible mechanism for the development of cancer is chromosomal damage, which may result from the heat generated in tissue by nonionizing radiation. Defects in the DNA can cause genetic mutations and can play an essential role in the transition of tumor growths from benign to malignant.

Because the biological effects of nonionizing radiation are given a high research priority in the Soviet Union, much work has been done there. And Russian scientists have been able to establish a relation-

A somewhat bizarre link between cancer and microwave radiation came to light at the height of the Cold War between the United States and the U.S.S.R. In 1962, it was discovered that the American Embassy in Moscow was being exposed to a continuous beam of microwave radiation as high as .018 milliwatts, almost twice the level of the Russian safety standard (but about five hundred times less than ours). The level can thus be interpreted as either injurious or harmless, depending on whose perspective we assume.

The State Department conducted a secret investigation called Project Pandora, and, when it was concluded that the radiation aimed at our embassy would cause no harmful effects, data used in the study were destroyed. The Russian action was presumed to be an attempt to deactivate electronic equipment on the roof of the embassy. At that time, a decision was made to suppress the fact that the building was being irradiated.

Embassy employees were officially told about the beam of microwave radiation in 1976, after the Russians increased its intensity. In the meantime, Ambassador Stoessel had become quite ill with a blood disease, nausea, and bleeding in the eyes. The furor over this bit of news prompted the State Department to conduct a massive study on some 4,000 former employees of the agency, and the final report reiterated that no health hazards had resulted from their exposure.

But remember the long latency periods of cancer. A few years later, Ambassador Stoessel's condition was diagnosed as either a lymphoma or a leukemia. Another chilling fact is that two of his predecessors, Charles Bohlen and Llewellyn Thompson, have already died of cancer. Coincidence? Perhaps.

ship between microwave radiation emitted in pulses and chemicals that cause mutations. The radiation alters cell walls and membranes, permitting easier entry of the mutation-forming chemicals and thus enhancing conditions that can lead to cancer. Other studies also suggest that the elevated temperatures caused by microwave radiation enhance the action of chemicals lethal to cells and thus contribute to their cancer-causing potential.

Studies involving large populations and exposure to radar or microwave frequencies would seem to support a link with cancer. For example, the World Health Organization has carried out studies in a small region of Finland adjacent to the Russian border and near an area where the Soviets had installed a warning radar facility. This system continuously exposed Russia's Finnish neighbors to high levels of ground waves and so-called scatter radiation. The study revealed an increase in the incidence of cardiovascular disorders, compounded by an increased incidence of cancer.

On a less global level, another possible link between cancer and microwaves was raised in a small New Jersey town. Five children at a particular elementary school in 1978 had developed cancer—an extremely unlikely event under normal circumstances. When officials began to investigate, they discovered more than 6,000 sources of microwave radiation within a fifteen-mile radius. And, sitting on top of a hill, the school was a perfect target for them. The National Bureau of Standards has measured microwave radiation in the area and found levels high enough to "imply something."

Five cases of cancer in one school might have been coincidence, but the odds against it, according to authorities, are about 10 million to 1. The occurrences are still being investigated. But the data available so far are not of sufficiently high quality for us to draw any definite conclusions about the cancer-producing potential of microwave or radar radiation. More quantitative data are needed.

DAMAGE DUE TO OTHER NONIONIZING RADIATION

ULTRAVIOLET LIGHT

Ample evidence implicates ultraviolet light, another form of nonionizing radiation, in biological damage to humans. Its effect on the eye, for example, is well documented. Even though they do not cause ionization, ultraviolet rays are energetic enough to produce excitation in eye tissue that can be very destructive. The most marked reactions to ultraviolet light occur in the cornea, conjunctiva, and lens of the eye. Mild reactions include inflammation of the cornea and conjunctivitis, or inflammation of the conjunctiva. More severe reactions, usually resulting from prolonged exposure, are seen in the lens, where cataracts can occur.

The other major organ system affected by ultraviolet radiation is the skin. An immediate effect, of course, is sunburn. But when exposure is prolonged, the results can be various skin eruptions popularly known as "sun poisoning," premature aging of the skin, and skin cancer. This most common form of cancer accounts for about 40 percent of all malignant lesions in the United States and about 2 percent of all cancer deaths, between 6,500 and 7,500 deaths in the United States every year.

Of particular concern to public health officials is the rising incidence of this form of cancer. A special study done by the National Cancer Institute revealed that the number of cases of skin cancer and deaths resulting from the disease had doubled in one large metropolitan area over a ten-year period. This rise in incidence appears to reflect the national average and has been attributed to the availability of more free time, which many individuals spend in the sun.

Extensive evidence leaves little doubt about the connection between ultraviolet radiation and skin cancer. For example, over 90 percent of skin cancers occur in parts of the body exposed to sunlight. We also know that the incidence of skin cancer in the United States increases as geographic latitude decreases: more cancer occurs in southern regions. More than 2.5 times as much skin cancer is seen in Dallas–Fort Worth as in Minneapolis–St. Paul.

Some people make better targets for skin cancer than others, because its incidence is affected by the amount of a person's skin pigmentation. Thus black people, whose skin is protected by pigmentation, develop far less skin cancer than white people. Fair-skinned, blue-eyed people seem especially susceptible, as attested to by the high rates of skin cancer in countries populated by people of Celtic origin, such as Australia and Ireland. About 50 percent of all cancer in Australia is skin cancer.

The mechanism by which skin cells develop cancer probably involves DNA. When nonionizing radiation in the form of ultraviolet rays strikes this genetic material, the excitation produced results in biological damage. If such damage to the DNA is not repaired, the cell can die, producing the appearance of premature aging of the skin. But ultraviolet radiation can initiate the cancer mechanism instead by allowing the cell to survive, crippled by errors in DNA function that produce mutations and ultimately malignancy. Most dermatologists advise against excessive sunbathing, because the possibility always exists that ultraviolet radiation will damage a particular part of the DNA and activate the cancer-forming mechanism.

INFRARED RADIATION

Another important cause of biological damage from nonionizing sources is infrared radiation. This, of course, comprises a section of the electromagnetic spectrum just above microwaves and just below visible light. Here the energy levels are such that, when radiation is absorbed, excessive heat is generated within the material by the vibration and rotation of its molecules.

Infrared sources may be either natural—the sun—or artificial—various kinds of heaters, diathermy devices, and illumination sources such as fluorescent, incandescent, and high-intensity lamps. Generally, biological damage due to infrared radiation is associated with radiant heat. When individuals are exposed to high levels of radiant heat in their work, a condition known as thermal stress may result. Thus bakers, cooks, chemists, firefighters, steel mill workers, and welders may experience weariness, weakness, or malaise. More often, however, infrared radiation affects a person's eyes or skin.

When an individual is exposed to infrared radiation, the first area to be affected is the cornea. This structure of the eye possesses nerves that can easily detect even small rises in temperature, and when heat enters the cornea, distinct pain is felt. Beyond the acute sensation of pain, prolonged exposure to infrared rays can produce a condition known as lens cataracts. This opacity is thought to be due directly to absorption of the infrared rays by the lens or indirectly to heating of fluids in the eye.

Injury to the retina in the form of burns can also occur, and exposure to infrared radiation causes increased evaporation of tear fluid in the eye, compounding a condition known as "dry eye" if it already exists.

In skin, the absorption of infrared rays depends on the amount of pigmentation present, blood pigments, and the manner in which rays are scattered by the structure of the skin. Among the somatic effects that infrared radiation can produce in skin are burns, dilatation of small arteries, and increased skin pigmentation. Infrared rays can also affect the delicate outer skin of the eyes, and exposure has been associated with a condition called blepharitis, or inflammation of the eyelids.

No link between infrared radiation and skin cancer has turned up, but the rays may work in concert with other agents that do cause cancer. A report published in the 1940s pointed to an increased incidence of cancer in people who were occupationally exposed to heat. However, these individuals were also exposed to other cancer-causing agents such as sooty deposits, tars, and ultraviolet rays. Any research on the cancer-causing potential of infrared radiation undertaken in the future should obviously include studies on the incidence of cancer among people who are occupationally exposed to excessive heat.

Beneath the surface of the skin, infrared radiation has been associated with impaired blood flow through the vessels of the spleen and kidneys. And the body's immune mechanisms, which work to keep individuals functioning normally, also seem to be affected. At low levels of intensity, the protective mechanisms seem to be stimulated, whereas higher levels of infrared radiation seem to reduce the body's ability to protect itself. The ability of nerves to transmit impulses can also be temporarily impaired when the nerve undergoes infrared radiation.

The tissues most sensitive to these rays are obviously the ones on or just below the skin. As might be expected, such tissues include the testicles, which are actually located outside the body and are very sensitive to heat injury. What occurs when the tissue is subject to radiant heat is a condition known as oligospermia, a deficiency of sperm cells. This disorder is generally temporary and appears to correct itself as the testicular tissue recovers from its heat injury.

Another biological effect of infrared irradiation is its slowing action on spontaneous chromosomal repair. If the rays can cause this effect, it is not unlikely that they can also produce mutations. However, no evidence has emerged to suggest that infrared radiation can cause hereditary changes.

VISIBLE LIGHT

Finally, we come to the subject of visible light, that portion of the electromagnetic spectrum which the human eye can see. Here, of course, the main source of trouble is lasers. In the late 1970s and early 1980s, the rise of the discothèque as a national institution markedly increased the use of lasers for entertainment purposes. In fact, the laser light show has become a standard form of entertainment. Visually, the effects can be spectacular. These exciting effects are produced by diffusing the concentrated beams of light and then optically refracting them for display on a screen or the dome of a planetarium. Usually accompanied by music, these beams can be made to undergo changes in slope and almost move with the rhythm of the music. But in order to make these shows more and more exciting, their producers are using laser beams of higher and higher power. These beams are biologically hazardous.

Striking a dull surface, the high-intensity laser beam is not dispersed very widely. But when it strikes a shiny surface, the beam can be reflected back at sufficient strength to cause retinal burn if it enters the eye. Blindness can even result if the beam enters the eye in just the right direction. How much damage a laser can cause depends on certain characteristics of the beam. For example, damage is greater when a particular quantity of energy is delivered over a short period of time, when the beam is emitted in repeated pulsations rather than as a single pulse over a specific time period, and when the wavelengths are shorter (frequency is greater and penetration into tissue deeper).

The exact mechanism by which laser light causes eye injury is not yet completely known, but its effects on tissue seem to be basically the same as those of ultraviolet and infrared radiation.

PROTECTIVE STANDARDS

We have seen in this chapter that nonionizing radiation, even from some completely unexpected sources, is capable of wreaking a considerable amount of biological havoc. Where the sources are natural, a few commonsense precautions will usually suffice to reduce any risk. (We will discuss risk reduction in the third part of this book.) Where sources of nonionizing radiation result from human activities, certain standards have been set in the United States for our protection.

The most significant of these are microwave protection standards, and they are of two types. Personal protection standards are related to exposure, and device performance regulations are related to emission of radiation. As we have noted, the personal occupational standard in the United States today is 10 milliwatts per square centimeter —a thousand times greater than the standard in the Soviet Union, where the feeling seems to prevail that *any* exposure to microwave radiation should be avoided. No specific standards exist in the United States for microwave exposure to the general public.

As for performance regulations, the standard in effect for home microwave ovens at a distance of 5 centimeters from the oven door is 1 milliwatt per square centimeter before the oven leaves the factory, and 5 milliwatts per square centimeter during the lifetime of the oven. A "leaking" microwave oven is dangerous only if one is close to it. The power level decreases rapidly as the distance increases. In addition, the U.S. government insists on such safety features as interlocks to prevent any generation of microwave power with the door open.

Unfortunately, very few facilities that repair microwave ovens are equipped with radiation monitors to detect and correct leakage problems. And surveys have shown that many ovens in operation have improper door seals or loose hinges that permit microwave radiation to leak out into the kitchen.

At present, there are no ultraviolet emitting standards. Exposure is largely limited to work settings where devices that produce ultraviolet radiation, such as arc welding equipment and carbon arc lamps, are used. The current allowable occupational exposure to ultraviolet radiation was established by the National Institute for Occupational Safety and Health in 1972. It is less than the amount a sunbather would receive in one or two hours on a sunny summer day at the beach.

At this writing, no official standard exists for exposure of the skin or eyes to infrared sources. Nor do threshold limits exist for occupational exposure of the cornea, lens, and retina. There is obviously a pressing need to determine an exposure standard for infrared radiation. As a working standard for infrared radiation, many scientists today project a figure of 10 milliwatts per square centimeter.

Because high-intensity laser beams can cause blindness, the FDA issued a safety standard that became effective in August of 1976. Laser products were divided into four categories on the basis of levels of radiation produced and anticipated biological damage. In Class I were products that produced radiation below levels that have been found to cause harmful effects. Class II lasers give off visible light and can cause injury to eye tissue after prolonged exposure. Class III includes devices with beams sufficiently high in intensity to produce biological damage to tissue after one direct exposure for a short time. And Class IV lasers are powerful enough to produce harmful effects not only by direct exposure, but by diffuse reflections as well.

The standard issued by the FDA also requires a variety of safety features, including protective housings, safety interlocks, key switches, emission indicators, beam controls, and warning labels. Which items on this laundry list of safety equipment are required depends on the laser product class. The higher the power, the greater the potential for biologic damage and the greater the number of safety devices that are required. Where lasers are used for special purposes, such as medicine, entertainment, and surveying, additional safety features have been mandated.

At present, lasers intended for classroom use or entertainment— so-called demonstration instruments—are required to meet FDA standards for Class I or Class II lasers. But with more powerful instruments being used for greater dramatic effects, the danger to audiences is growing. Light shows currently are subject to little or no local regulation. While twenty-three states have laws on the books that govern nonionizing radiation, very few have specific statutes covering lasers. Two states that do have shut down unsafe light shows. But monitoring these shows is almost impossible, because most of them play one-night engagements and are gone before an inspection can be made.

There is no doubt that nonionizing radiation—radar, microwaves, infrared rays, ultraviolet rays, visible light, and lasers—can do serious damage, but these waves, rays and beams have also added immeasurably to the quality of our life. A continuing dilemma will be how to take advantage of the wonders inherent in devices that emit radiation . . . without paying a price in health status.

The average American does not know how to do this. In too many cases, neither do the scientists. Much work remains to be done if we are to prevail in the environments we create. But the game is worth the candle, as the following chapter will explain.

8
The Up Side

In the previous two chapters, we have considered the negative aspects of radiation—the human organism put in harm's way by the fearsome potential of radioactivity. But what is the alternative? Though we may reminisce fondly about the day when exposure to radioactivity was almost always natural, we can't go back. Nor would we want to.

But in the wake of a crisis such as that which occurred at Three Mile Island, sentiment begins to grow that radiation is too dangerous and that nuclear reactors should be outlawed. A "spillover" antiradioactivity effect is also felt in other areas, and medical uses of radioactivity are questioned. However, before consigning all devices that involve radioactivity to the scrap heap, we must weigh their risks against their benefits.

This is something most people do constantly. Every time one chooses to go on an airplane or automobile trip, one decides that the benefits outweigh the possibility of an accident. Deciding whether to invest in the stock market involves analysis of the relative likelihood of profit and loss. And when a proposed shopping mall is approved in a community, the increase in business it will bring is deemed to outweigh a possible increase in traffic accidents. In order to make these judgments, we must know the benefits as well as the risks—the good news as well as the bad.

MEDICAL DIAGNOSIS

X-RAYS

In the area of medicine, the lives of untold thousands of individuals are saved or prolonged and the continued health of millions preserved every year by medical techniques involving a wide variety of radiation-producing devices and materials used to detect and treat disease. Among these are *diagnostic x-rays*. Since the x-ray was first discovered, x-rays have been used to detect abnormalities deep inside the body where direct physical examination is impossible. Fractured bones, kidney stones, and tumors have been made "visible." Techniques have also been developed by which patients ingest inert, heavy substances (such as barium) that absorb x-rays and therefore show the outline of intestinal structure by contrast. As a so-called contrast medium, barium can reveal the presence of such pathologic conditions as tumors in the gastrointestinal tract, ulcerations in the stomach, small bowel, and large bowel, and other abnormalities that can disable or kill.

X-rays can point a diagnostic finger at disease of the kidney, urinary tract, and lung and at various abnormalities of the heart. Films can pinpoint the location of neoplasms benign or malignant and of polyps and cysts. They can confirm the presence of bone diseases, tuberculosis, and joint diseases such as arthritis. Often a single film is not sufficient for diagnosis, because it is necessary to visualize movements or change. In such cases, doctors use the *fluoroscope*. This device takes a series of x-ray pictures in much the same way that a motion picture camera takes conventional pictures. The x-rays are then projected on a screen to provide a moving picture of the affected area. This picture can also be recorded on film or videotape for later study. And newer instruments come with special devices that produce better images with less exposure.

Dentists, orthodontists, periodontists, and oral surgeons use x-rays to detect decay between teeth, gum disorders, abscesses, impactions, and fractures or cancer of the jawbone.

When x-rays are used to confirm the presence of a suspected disorder, such as a broken bone or an ulcer, the benefits are obvious. X-ray use is more controversial, however, when it is applied for mass screening purposes, to detect disease before symptoms appear, and hence to improve the chances of effecting a cure or arresting the disease early.

Among mass screening programs, the chest x-ray has been particularly widespread. At one time, such examinations were routinely administered for the purpose of discovering the presence of tuberculosis. The incidence of this disease has declined to such an extent in recent years that mass screening by x-ray has been largely discontinued on the basis that the attendant risk, however small, outweighs its benefits. A totally different case is *mammography*—the widespread practice of x-raying the chest area in order to detect breast cancer.

Mammography has become controversial in a very short time. It was the subject of nationwide publicity in 1974 when, through its use, breast cancer was discovered in the wives of both the president and the vice-president of the United States. Both women underwent breast surgery and both survived. Stirred to action by those developments, women rushed to their doctors demanding mammographic examination—a positive step that can definitely save lives.

In the United States today, statistics indicate that about 7 women out of 100 will develop breast cancer, the most common cancer found in women and the leading cause of cancer death in women. About 70,000 new cases of breast cancer are diagnosed annually, and each year almost 30,000 women die of the disease. Mortality can be reduced when the cancer is discovered early, before it has firmly established itself, has had a chance to grow, or has metastasized (spread to surrounding tissues). In the early stages, breast cancer symptoms may not be present. But if breast cancers can be detected before symptoms appear, the chances of a cure are that much greater. Earlier discovery, therefore, is the fundamental premise of screening programs.

And these programs work. In studies conducted by the National Cancer Institute and the American Cancer Society, involving approximately 250,000 women, mammography was found to be effective in the early detection of breast cancer in both younger and older women. About one-third of the cancers discovered were in early stages and, in more than 70 percent of the cases, the cancer had not spread beyond the breast.

The value of mammography in detecting cancer has been clearly demonstrated. Figures vary on the relative efficacy of physical examination and mammography. What does seem clear, however, is that using both techniques together improves the physician's chances of detecting breast cancer.

According to recent surveys, radiation exposure occasionally varies among hospitals and clinics that perform mammography examinations. And improper technique or faulty equipment may also expose some women to unnecessary levels of radiation. These facts have sometimes been given disproportionate publicity in the media, where the risks have been emphasized and the benefits minimized. With opinion divided even among experts, the tendency has been to rule out annual mammography in women under fifty with no complaints or physical symptoms and no history of breast cancer in their mother or sisters. For women without symptoms, the first mammographic examination should be performed between the ages of thirty-five and fifty, but many authorities now feel that routine annual mammography screening should not start until the age of fifty or older, when the chances of reducing mortality are greatest.

But according to the National Cancer Institute, times have changed. As a result of improvements in technique, women are now being exposed to considerably lower radiation levels during mammographic examination. With these low doses, the risk from mammography may be essentially negligible when compared with the benefits that would be derived from early detection of breast cancer.

Approximately 40 million women in the United States are currently at risk for breast cancer. It has been estimated that use of the present low-dose technique over a thirty-year period could reduce the annual number of deaths due to breast cancer by one-third—a total of 12,000 women saved each year.

CAT SCAN

When a patient is ill and in need of decisive action on the part of his or her physician, the risk of developing cancer (about 1 in 100,000) some ten to thirty years later seems like a small one. If benefits to the patient are real and immediate, the small risk of cancer may be acceptable. This is the philosophy underlying a new diagnostic x-ray technique that can expose patients to relatively lower doses of ionizing radiation risk than would have been received in other diagnostic x-ray procedures.

This new type of x-ray system combines x-ray equipment with a computer to produce images of cross sections, or thin slices, of any region of the body as they would be seen looking down through the patient from head to foot. The procedure is called *computerized axial tomography,* or simply CAT, and derives from the Greek word *tomos,* which means "cutting or slicing." Images, or scans, produced by this technique allow precise visualization of abnormalitites that would not be seen at all or would appear fuzzy and indistinct in conventional x-ray films. And the procedure is fast, simple for the patient, and painless.

CAT scanning overcomes two basic drawbacks of conventional x-rays. In the ordinary x-ray picture, images of objects at different depths within the body overlap and therefore obscure each other's outer edges on the film. Too, if adjacent tissues are of like density, they absorb the same amount of radiation and x-rays are unable to distinguish between them. Thus a tumor and the tissue surrounding it may be indistinguishable. The problem is usually "solved" by taking a large number of pictures from many angles—and exposing the patient to large doses of radiation in the process.

CAT scanning largely eliminates these problems. X-rays are taken in such a way that only one plane through the body, usually about 1 centimeter thick, is seen in sharp focus, while the other planes are blurred. The scan is made possible by the computer, which sorts through as many as 1,000 or more views and reconstructs them into a single image. The first computerized tomography machines, called "head scanners," were designed only for study of the brain within the skull. But shortly after their introduction in the early 1970s, "body scanners" were made available. Studies done in 1977 indicated that accuracy of diagnosis with these scanners ranged from 80 to 100 percent.

The CAT scanning technique is especially useful for visualizing abnormalities of soft tissues in the brain. Using this device, doctors can now distinguish between different types of strokes, detect whether organic brain damage is the cause of certain types of insanity, and visualize lesions of the optic nerve. In the body, CAT scans can reveal large tumors in the liver, pancreas, kidney, and other structures; these tumors would be invisible with conventional x-rays. Projected applications include differentiating between such abnormalities as cysts, abscesses, and tumors; determining whether tumors are benign or malignant; monitoring complications that may follow surgery; and checking on the results of cancer therapy.

Another important advantage is that the computer component of the CAT scanner can store information, thus becoming in itself an aid to diagnosis. And, in terms of the patient, the procedure is without side effects and economically feasible. CAT scanners can be used for patients in the hospital and outside. They can reduce the number of hospital admissions and shorten the length of a hospital stay. For these reasons, CAT scanners have been hailed as one of the outstanding medical innovations of the 1970s.

Nor is the x-ray the only form of radiation that is applied diagnostically. In medical research, microwave radiation is being tested as a diagnostic tool for warming blood and other tissues to retard tumors and to promote the healing of skin wounds. Microwave radar has been tested for monitoring movement in the walls of arteries, and it may provide clues to the origin of such cardiovascular diseases as stroke and coronary artery occlusion (heart attack).

MEDICAL TREATMENT

Aside from being useful as an aid in diagnosis, radiation has tremendous value as a therapeutic measure, to help cure or manage an illness that has already been diagnosed. Perhaps its most widely used therapeutic application is in the treatment of cancer.

As a cause of death in the United States today, cancer comes right behind heart disease and accounts for almost 17 percent of our annual mortality. The American Cancer Society estimates that this represents approximately 1,000 people per day. Nevertheless, a diagnosis of cancer is not necessarily a death sentence.

Authorities estimate that 33 of every 100 patients with cancer will be cured, and perhaps 17 more can be cured if diagnosis is made early enough. In other words, half of all cancer patients need not die of cancer. The number of living Americans who have been cured of cancer may today exceed 2 million. And many of them owe their lives to ionizing radiation. About 50 percent of all cancer patients undergo radiation therapy seeking a cure or the relief of symptoms.

X-RAY THERAPY

Radiation therapy is predicated on the ability of ionizing radiation to make cells incapable of reproduction, thus effectively destroying them. Cancer grows by the rampant proliferation of diseased cells. Thus any agent that arrests this uncontrolled cell division controls the growth of cancer.

This works fine in theory, but it is not always so easy to hit the apple *and* miss William Tell. Obviously the radiation source, usually an x-ray beam, must be aimed very accurately to affect only the cancerous area. But it is almost impossible to keep normal tissue from being exposed to some degree of radiation. For this reason, the quantity of radiation used to treat a particular form of cancer depends not on the tumor, but on the ability of surrounding healthy tissue to withstand that radiation. This level is critical, because the difference between the radiation level that can kill cancer cells and that which can kill normal cells is often very small. If not precisely determined and administered, a radiation dose may produce effects that are worse than the disease it was designed to cure.

One way out of this quandary is to administer small doses on a daily basis for extended periods of time. Experience has shown that this technique increases the chances of successfully destroying malignant cells while minimizing damage to healthy tissue.

Among the types of cancer that are especially amenable to radiation therapy are cancer of the cervix, cancer of the larynx, certain types of bone tumors, and Hodgkin's disease, a form of cancer of lymph tissue. Radiation therapy has also shown promising results in certain kinds of skin cancer and in seminoma, a cancer of the testicles. In the latter two conditions, radiation therapy is often part of a therapeutic program that also includes surgery.

Radiation can be transmitted to the target areas by several means. The most widely used are x-ray beams generated by machines capable of operating at extremely high voltages to increase penetrating power. New linear accelerators, for example, can operate at up to 8 million volts and reach even the deepest tumors. Machines that use cobalt-60, a radioactive isotope that gives off extremely energetic gamma rays, are another source of penetrating radiation. Gamma rays produce biological effects similar to those of x-rays. Because cobalt units require less electricity and infrequent servicing, they are particularly suitable for hospitals in smaller communities.

Other devices generate neutrons or microwaves instead of x-rays or gamma rays. Temperatures ranging from 107.5° to 109.5° Fahrenheit seem capable of destroying cancer cells while sparing normal tissue. And, as we have seen, microwaves can generate such temperatures within tissue. Melanoma in particular seems to respond well to the hyperthermic levels generated by microwaves.

RADIONUCLIDES

Radiation treatment may also be carried out through a technique known as *brachytherapy*, which involves implantation of sealed radionucludes directly into the cancerous areas of the body. This is often the method used in treating cervical cancer and cancer of the prostate. Other conditions treated with radionuclides include thyroid disorders and thyroid cancer.

Procedures of this type are part of a new and rapidly expanding medical specialty, *nuclear medicine.* A relatively recent development in the medical application of radioactivity, nuclear medicine dates back only to 1946. The intervening years have seen the introduction of new and improved pharmaceutical radionuclides as well as increasingly sophisticated instruments.

Radiopharmaceuticals give off radiation that can be identified and measured within the body (see Chapter 4). This enables physicians to keep track of body physiology and, through the detection of abnormalities, to arrive at diagnoses. For example, one of the most widely used radionuclides for this purpose is technetium-99, which, when combined with a pharmaceutical product, will gravitate to specific parts of the body. There its metabolism can reveal specific defects. Colloidal sulfur tagged with technetium-99 is now used extensively for examining, or scanning, the liver, spleen, and bone marrow.

New diagnostic tests have recently been developed that involve the use of radioisotopes in the laboratory rather than in the human body. Biological materials from the patient's body are used in tests of thyroid function, for example, eliminating the need to expose the patient to radiation.

Another fascinating application of radionuclides is based on radioactive decay as a small power source in the range of 1 to 100 watts. Such radioactive units are used as power sources for mechanical hearts and certain cardiac pacemakers.

OTHER RADIATION THERAPY

The laser is a further application of radiation in the field of medicine. Intense laser light is used in surgery for correction of a detached retina in the eye. The retina is reattached through the laser's ability to coagulate blood. Lasers have also been used in the treatment of skin angiomas. At lower energy levels, laser beams can selectively alter tumor cells in skin, thus providing an additional therapeutic means to combat skin cancer. Higher-energy beams can be used almost as a surgical scalpel to remove tumors from deeper tissues. And laser physics is also used in general surgery and neurosurgery.

Because ultraviolet rays reduce the ability of cells to support the growth of specific viruses, they point the way to possible cures for many of the infectious diseases that now plague us. Visible-light radiation therapy has been found to help in the treatment of hyper-bilirubinemia, a jaundice of the newborn, which afflicts about 35,000 newborn infants each year. And laser radiation is under intensive investigation in as-yet-experimental ophthalmic surgical procedures. It can be aimed precisely, and such surgery would be bloodless.

In the area of dentistry, ultraviolet radiation has been put to some interesting uses. Superior new dental cements, which greatly expand the dentist's options in restoring or maintaining good dental health, can be hardened only under ultraviolet radiation. Dentists can also reveal the presence of plaque on teeth by applying a material that adheres only to plaque and gives off a characteristic glow under ultraviolet light. And photography inside the mouth is made possible by ultraviolet light.

Infrared light is the basis for a diagnostic procedure called thermography, which measures the infrared radiation emitted by skin and provides a pictorial representation on a display panel. It is used widely to detect breast abnormalities. If cancer is present, the thermograms for the cancerous breast and the healthy breast usually differ in appearance. The advantages of this procedure are that it does not involve surgical invasion of the tissue as a biopsy would and that it eliminates radiation risk and costs less than mammography. Thermography is not yet so accurate as mammography, but the procedure may be refined through technological improvements.

NUCLEAR ENERGY

Aside from medical use, radiation is also useful in the production of energy, a point that even critics of nuclear power concede. It is a matter of record that the United States, the citizens of which number only about one-twentieth of the world's population, produces and uses more than one-third of all the electricity produced in the world.

What's more, between now and the end of this century, the United States will probably use as much fuel to produce energy as we used in all our previous history.

Obviously, our own supply of conventional fuels will run out, leaving us at the mercy of foreign oil-producing powers who will be able to demand—and get—exorbitant prices for their product. And in the meantime, the coal and other fossil fuels we use to generate energy will darken our skies, pollute the air we breathe, and poison our streams.

This is *not* to say that nuclear power is the panacea we all seek. Production of such power is not without its own set of problems in terms of the environment, the health of the community, and even the safety of citizens in large geographic areas, as the events at Three Mile Island made abundantly clear. But that insistent question keeps bobbing to the surface: What is the alternative? At present, no practical alternatives to fossil fuels or nuclear power exist for generating the quantities of power needed by an industrial society.

Perhaps the main benefit nuclear power bestows is that it is not produced by the burning of fossil fuels, which adds sulfuric acid and sulfur dioxide to the air. Coal not only contributes heavily to air pollution, but it is also very dangerous to obtain; accidents in coal mines are common, and there is no way of telling how many men have lost their lives deep in the earth. Millions have contracted black lung disease, the prime occupational hazard of coal mining. And it takes tremendous quantities of coal to match the energy produced in a nuclear power plant from infinitesimal quantities of uranium. As for oil, the use of this rapidly dwindling resource to generate electrical power will soon leave insufficient quantities for essential (not to mention recreational) transportation.

The advantage of nuclear power is the relative ease with which it can be obtained, with minimal disruption of the environment or pollution of the air. Despite a few mishaps (see Chapter 5), the nuclear industry in the United States can boast of a remarkable safety record in terms of the number of directly involved individuals killed or injured on the job. Statistics bear out that generating electricity via nuclear power is far less expensive, in human suffering and loss of life, than producing it from coal, the mining of which kills and disables so many.

According to a report filed by the Congressional Office of Technological Assessment in 1979, our use of coal at present-day levels may be responsible for an estimated 48,000 premature deaths a year. The figure will rise to 56,000 a year by 1990. And if we go back to coal as a replacement for nuclear power, it will climb even higher.

Partly as a result of vociferous opposition to nuclear power plants, a great deal of attention has been focused on the safety of the fission plants and fusion reactors just coming into experimental use. Regulatory agencies have lowered allowable levels of radioactivity released from nuclear power plants, and their scrutiny of all phases of reactor operation is painstakingly thorough.

Gamblers at a Las Vegas gaming table risk losing their money only because they perceive that they have a good chance of winning. But without possible benefit, no risk is acceptable. In the area of energy, every means of production involves some risk to human life. We must obviously take a long, hard look at the benefits of nuclear energy and weigh them against the risks, because one fact has become unmistakably clear: If we are to survive as a nation, we must have some alternatives ready when our supplies of oil are exhausted.

MILITARY AND CIVILIAN APPLICATIONS

The future of the United States as a nation may also depend on our ability to defend ourselves. And a major deterrent to aggression by a potential enemy is our nuclear arsenal. Like the man who "rode shotgun" on the Old West stagecoaches, sitting next to the driver with his weapon in full view across his knees, nuclear capability tends to preserve the peace by its mere presence.

The concept of nuclear weaponry as an olive branch may not sit well with pacifists, but the attempts that the leading nuclear powers have made to limit strategic arms have been based on an exquisite understanding of what could happen in a nuclear war. The question of which nation owns more weapons is really academic. Whether the United States has 5 or 10 percent fewer or more makes no difference, because, with what we can already deploy, we could wipe out any potential enemy many times over. With no chance to win, the likelihood that any nation would attack us is practically nonexistent. In this sense, nuclear weaponry provides the feeling of security we need to pursue positive national purposes.

Radiation also turns out to be an exceptionally good servant in other areas of our national defense effort. Atomic-powered submarines of the U.S. Navy can circumnavigate the globe several times with just one load of fuel. Radar frequencies are used in missile guidance systems to aim ground artillery, to give advance warning of any attack launched toward our shores or borders, and as navigational devices on our strategic bombers and other military aircraft.

Certain other uses of radiation provide benefits in both the military and civilian sectors. In telecommunications, for example, extremely powerful transmitters that can scatter microwaves through the troposphere have been set up in the United States and abroad as links for satellite and nonmilitary communications systems. Two microwave broadcasting satellites in fixed orbits above the earth since 1979 relay telephone conversations, TV and radio broadcasts, and printed materials to receiving stations throughout the United States. Today, hundreds of microwave telephone and TV signal relay towers are strung across the country.

Other such towers provide links for motorist aid call boxes along the nation's highways. And private microwave relay systems have been installed, so that computers in different cities can communicate with each other. The millions and millions of citizen's band radio transmitters and all the "good buddies" who operate them broadcast on microwave frequencies from their cars, trucks, or residences. The availability of microwaves and high-frequency radio waves has made it possible for people to communicate with each other as never before.

In recent years, lasers have also come into use in communications. A number of laser detection and ranging systems are already in use (ladar—by contrast with radio detection and ranging, which we know as radar). It is theoretically possible for one high-frequency laser signal to carry 10 million simultaneous phone calls or 8,000 simultaneous TV programs. Such is the mind-boggling potential of lasers in communications.

The list of positive contributions that radiation devices can make is almost limitless. Radiation can be used in the prevention of food spoilage; it sterilizes by killing all bacteria, microorganisms, and spores. High doses are necessary for some types of bacteria, such as the lethal varieties that cause food poisoning, and irradiation produces heat within the food that may begin to cook it. The technique is effective for large quantities of foods, however. Additional research is underway to explore and expand the possible uses of radiation in food preservation.

Radiation in the form of radar is used in meteorology to detect weather disturbances and spot clouds with the type of vertical development that causes thunderstorms, thus making it possible to warn aircraft away from these areas.

Radar also helps merchants keep the price of their merchandise down by reducing pilferage. In many stores today, items are tagged with radar-detectable markers that are removed at the time of purchase. When a shoplifter tries to leave the store with the tag still attached, a radar scanning system sets off an alarm.

Anyone who has traveled by air during the last few years is familiar with the use of x-ray systems as a means of inspecting luggage at airports. From 1960 to 1972, 147 attempts were made to hijack U.S. airplanes, and 91 were perpetrated by individuals carrying concealed weapons. In 1969 alone, 40 attempts took place and 33 were successful. The present program was ordered by the president in early 1973, and commercial airlines began screening passengers and baggage for weapons. Skyjacking has virtually disappeared. The price is a microdose of radiation.

So many of the conveniences we enjoy today are made possible by radiation devices that listing them would be next to impossible. Included would be everything from microwave ovens to garage door openers to TV remote-control devices. And radiation technology is increasingly being applied to meet the growing needs of commerce, industry, and science.

At the punch of a few buttons, video display panels in banks, airline reservations offices, and other commercial establishments can reproduce portions of the astronomical numbers of data stored in computers. The price of the time and labor saved is just a trace of user exposure to radioactivity; video display units give off no more radiation per hour than normal background levels.

Another intriguing application of radioactivity is in activation analysis—a technique used to identify the chemical composition of various substances. Through irradiation, trace elements within the sample are made temporarily radioactive. Characteristic patterns of ray emission can be used to identify these elements. The technique has been used in archeology to pinpoint the place of origin and age of various artifacts, in forensic medicine as an aid in crime detection, and as a research tool in biochemistry. Activation analysis was also used to identify some 70 percent of the elements in rocks brought back from the moon by U.S. astronauts.

In industry, applications of radioactivity abound. And in agriculture, radiation makes important contributions that help increase crop yield and keep prices down. Using a gamma ray source and a counter, the agricultural technician can determine how compact the soil is by first measuring the absorption of neutrons by both soil and water and then comparing it with their absorption by water alone.

Radiation is also used in insect control. Large numbers of the offending insects are bred in captivity, irradiated with gamma rays to render them sterile, and then released to mate with nonirradiated insects. Obviously no offspring are produced, and the population of the insect pests gradually diminishes. An advantage of this method is that birds and other predators on insects are not poisoned, as they are by insecticides. In the United States, Spain, Italy, and Israel, this method has been used to eradicate the Mediterranean fruit fly. And it is now being tested on so-called killer bees, a particularly aggressive and harmful strain presently migrating northward from South America.

The list could go on endlessly, and it may well extend into new and exotic applications that open exciting vistas on earth and in the vast reaches of space. But anyone who wants to dance must pay the piper. There is only one consideration: The value we attribute to the dance must exceed the cost of the music. One very good way to increase benefits is to reduce the risks. Reducing those risks is the subject of the third part of this book.

part three

Reducing
the Risks

9
Making the Use of
Ionizing Radiation Safer

Of the sources of ionizing radiation discussed in Chapter 4, the cosmic and terrestrial varieties leave very few options to people who want to minimize the risks of exposure. They might decide to live at sea level rather than in the mountains or to take fewer airplane trips, but the net reduction in exposure would make very little difference, if any, in their future health or life span. Natural ionizing radiation is a fact of life that is usually beyond our control.

Radiation that results from human activities is another matter. Of the two most prominent sources—the nuclear power industry and medical and dental facilities—the former is by far the less important. Despite all the publicity about Three Mile Island, the fact remains that radiation leakage resulting from this unprecedented near catastrophe can be expected to cause just *one* fatal cancer more than the 350,000 that would normally be expected among the 2 million local residents. (Two different government agencies have made this estimate.)

Other authorities estimate that individuals living within five miles of a nuclear reactor increase their risk of contracting fatal cancer by about 1 part in a million. According to one expert, this entails roughly the same risk as smoking 1.4 cigarettes, drinking half a liter of wine, eating 100 charcoal-broiled steaks, living two months in a stone or brick building, or breathing the air in New York City for a couple of days. When functioning normally, which is practically always, a nuclear reactor poses very little risk to life or limb.

Medical and dental x-rays pose significantly more risk, but not so much more that a patient should refuse an x-ray examination that the doctor deems necessary. Weighing benefit against risk, we find little likelihood that any patient will develop cancer as a result of an x-ray

procedure. At the same time, it is very likely that serious consequences will result from failure to have the x-ray done. The potentially risky x-rays are those that are *not* medically indicated and
provide no benefit to the patient. These, of course, should be avoided.

WHY SO MANY UNNECESSARY X-RAYS?

One of the primary reasons why so many x-rays are taken unnecessarily is poor clinical judgment on the part of the physician or hospital
policy makers. In some hospitals, every patient admitted receives a
chest x-ray *routinely,* regardless of age, state of health, or length of
time since the last chest x-ray. Some x-rays are ordered out of habit,
and some are ordered in response to peer and patient pressure not to
miss any possible abnormalities, no matter how insignificant. Occasionally x-rays are taken to further medical education, even though
a radiologic examination is not medically indicated.

X-rays are sometimes used simply to placate patients and demonstrate that something positive is being done to improve their condition. This tendency is probably encouraged by medical insurance
programs that make such x-rays possible at no cost to the patient. Too
often patients switching hospitals or doctors are subjected to repeat
x-rays, because the second facility or physician doesn't trust the work
of the first or because repeating is easier than obtaining the previous
set.

Another reason why x-rays may be taken unnecessarily involves
what is known as defensive medicine—taking whatever measures are
necessary to avoid malpractice suits. As patients are awarded larger
and larger sums of money in malpractice suits, physicians' decisions
about patient care and diagnostic work-up are increasingly governed
by legal considerations.

X-ray examinations are often carried out for the legal protection
of employers. Routine pre-employment x-ray examination of the
lower back before hiring men who will be doing heavy labor is particularly widespread. For some civil service positions, pre-employment
chest x-rays are required. And some employers demand that their
employees submit to yearly chest x-rays or face dismissal. Many companies underwrite the cost of yearly physical examinations for their
executives; these examinations invariably include a chest x-ray and
may also include a barium enema study of the lower intestine.

Another major cause of unnecessary exposure to x-rays is the use of faulty equipment and poor technique on the part of technicians. For many years, the van housing a mobile x-ray unit was a familiar sight on city street corners and shopping center parking lots. Many of these mass screening programs were conducted by private or philanthropic organizations who were completely unaware that their improper use and maintenance of x-ray equipment was exposing innocent people to unnecessarily high radiation levels. According to a report issued in June of 1979 by the Department of Health, Education and Welfare, it is not unusual for radiation doses to vary considerably from facility to facility. Patients in one clinic or hospital can receive more than 100 times as much radiation per procedure as patients undergoing the same procedure in another facility. It just depends on how the x-ray unit is functioning.

Faulty equipment also produces poor pictures that have to be retaken, exposing the patient to twice as much radiation. This puts the patient in double jeopardy: not only is the radiation exposure increased, but an abnormality may escape diagnosis because of the poor quality of the x-ray pictures.

Where the cause of unnecessary x-ray exposure is purely mechanical (faulty equipment), the federal government has moved to protect the consumer and reduce the risk. Performance standards for x-ray equipment were set by the U.S. Food and Drug Administration in 1974. Adherence to these standards is enforced by the Food and Drug Administration. Its personnel review manufacturer reports, inspect manufacturing facilities, conduct laboratory tests, and field-test x-ray equipment. Any equipment that does not function properly or fails to meet performance standards during extensive testing is subject to immediate recall, by law.

While the teeth in this law are sharp enough to assure its enforcement, there just aren't enough inspectors to ferret out all equipment that does not comply with federal performance standards. Accordingly, pressure is now building to get individual states involved in inspecting x-ray machines at specific time intervals determined by the use and complexity of the equipment.

Many states now use an inexpensive means of measuring radiation that is actually a postcard. These cards are exposed to the various x-ray machines by technicians in hospitals and medical offices and sent back to a designated state office. Here they are "read" by special measuring devices that quickly reveal whether the equipment tested was functioning normally. In states where this program is in effect, radiation exposure has been reduced by at least an estimated 10 percent.

Other programs aimed at reducing the risk of overexposure to x-rays are also being investigated at the state level. Possible measures include guidelines for x-ray exposure, based on type of examination and type of machine; licensing procedures that provide for the periodic recertification and additional training of medical radiation technologists; and continuing medical education programs to cover the operation of all medical radiation equipment.

These steps are fine, as far as they go. But state and federal agencies do not have the authority to regulate how x-ray machines are used. Their efforts are limited to reducing x-ray risk in mechanical or nonjudgmental areas. Unfortunately, the issues surrounding medical x-ray examinations are not that black and white. When a physician tells a lay person that an x-ray is needed, on what medical grounds can that person dispute the recommendation? The reason why people go to doctors in the first place is because they are incapable of evaluating their own health status or treating their own illnesses. Looked at another way, why take your malfunctioning car to a mechanic if you're not going to follow her or his advice?

HOW CAN YOU AVOID UNNECESSARY X-RAYS?

Is there any way to make sure you receive an x-ray examination only when you need it? This is indeed a dilemma. But here are some practical tips that will help reduce your risk and still keep you on speaking terms with your doctor.

Don't make any unilateral decision to have an x-ray taken as part of a mass screening program. Many mobile units have x-ray equipment that undergoes rough handling and often exposes patients to excessive radiation. The mass tuberculosis screening programs that were once so prevalent are now frowned on by both the FDA and the medical community. TB is no longer a major threat to the health of U.S. citizens, and there are safer and more effective ways to diagnose tuberculosis. If you are healthy, you don't need a chest x-ray as a precautionary measure. Have x-ray examinations done only on the advice of a physician.

Don't insist that your doctor give you an x-ray as part of your examination. Patients very often demand radiologic procedures, because they are convinced that no thorough medical examination is complete without one. And physicians may agree in order to make the patient feel better psychologically and to avoid jeopardizing the doctor–patient relationship. If your doctor thinks you need an x-ray, he or she will suggest one. Otherwise, don't push for one. Ordering unnecessary diagnostic tests does not make a better doctor.

Make sure your doctor knows about any previous x-rays. The total *yearly* exposure is important, so your medical doctor should be made aware of the dental x-rays you have had, any pre-employment chest or spinal x-rays taken, and any x-rays that your previous doctor ordered. Most physicians are not reluctant to use each other's x-rays, and that can save you unnecessary exposure. You may want to start keeping a record of your own x-ray history, so that your doctor or dentist can refer to it whenever the question of obtaining an x-ray arises. Simply list the date of each x-ray exposure, the name of the physician or dentist who performed it, the type of examination, and its purpose. If you want to make this record as complete as possible, ask the doctor or dentist to list any special details of the procedure, such as shielding, filters, or the like.

If you are a female and think you are pregnant, make sure your doctor knows. An orthopedic specialist may not know what you and your obstetrician know. By all means, tell him. If the x-ray beams fall across a fetus, abnormalities may result. To avoid even the remotest possibility of such a tragedy, your doctor may elect to postpone the x-ray until after you give birth. Of course, if your doctor knows that you are pregnant and tells you that an x-ray is still needed, you should have it done. If you are planning to become pregnant, it is a good idea to have physical and dental examinations completed before conception.

Question a physician carefully about the need to x-ray your child. Children are especially prone to adverse effects from ionizing radiation. An expression of your concern may convince the physician not to x-ray if it is not absolutely necessary. On the other hand, if he or she insists that an x-ray is necessary, don't oppose it.

If you are a male, remember that your reproductive organs are especially vulnerable. You should be sure that your organs of reproduction are not exposed to direct radiation, especially if you contemplate becoming a father. X-rays can sterilize you temporarily or permanently, making parenthood impossible. Or, instead of killing sperm cells, the radiation may damage cells and raise the possibility of genetic problems in your children. One way around this problem is to insist that a lead shield or apron be placed over your reproductive organs, thus preventing any ionizing radiation from reaching them.

If an x-ray is necessary, try to have it done where a full-time radiologist is in charge. Experience has shown that patients generally receive less radiation exposure at such a facility than in a nonradiologist physician's office.

FURTHER SPECIFIC RECOMMENDATIONS

These general rules can minimize your exposure to x-rays significantly. Now here are some guidelines to help you reduce the risk in specific medical or dental situations that expose patients to ionizing radiation.

THE MEDICAL SETTING

One of these is routine screening for breast cancer through mammography. The very real benefits of this technique were discussed in the previous chapter, but many authorities feel that mammography should not be prescribed as a routine annual screening procedure in women under the age of 50, unless the patient or members of the patient's immediate family have developed breast cancer, or physical symptoms suggest the presence of the disease.

In general, it is a good idea to discuss all problems relating to your health with your own physician, because she or he knows your medical and family history. She or he may show you how to examine yourself for signs of breast cancer or suggest regular screening that does not include a mammography. And if you are going to undergo a mammographic x-ray study, do so in the best facility you can find, with the most up-to-date, low-dose mammographic equipment available.

Fluoroscopy is an extremely useful procedure, because it allows your doctor to observe your body systems *in actual operation*. But fluoroscopy also exposes patients to much more radiation than routine x-rays because they are under the direct rays of the fluoroscopic radiation for a longer time. Even the newer instruments that come with devices that intensify the image, and thereby reduce x-radiation exposure, deliver higher doses than standard x-ray machines. It would not be inappropriate for you to ask your physician whether fluoroscopy is really needed, as opposed to a conventional x-ray. Also ask whether the machine that would be used is equipped with an image-intensification device.

A doctor who suspects the presence of a tumor may suggest a radioisotope scan, a procedure that involves introducing a radionuclide into the body and measuring where it settles. Certain tumors absorb radionuclides at a different rate from normal tissue and are thus revealed as a brighter or fainter spot on a scan. The picture actually resembles an x-ray, and indeed the risks involved in a radioisotope scan are essentially those of an x-ray. To help reduce risk in a nuclear scan, follow the same general rules. But also remember that such scans can provide extremely valuable diagnostic information in cases where x-rays fall short. If your doctor thinks you need one, don't resist it.

Similarly, you should not resist computerized axial tomography. (The advantages of the CAT scan are outlined in Chapter 8.) If your doctor feels you need a CAT scan, try to find out whether a newer-model scanner will be used. These machines not only help overcome problems caused by the motion involved in breathing and the heartbeat, but they also cut down on your exposure to radiation.

THE DENTAL SETTING

Perhaps the most common x-ray exposure among Americans is the one experienced in a dental office. It is not uncommon for dentists to take a complete set of x-ray films routinely for every patient, during regular semiannual examinations. Anywhere from twelve to eighteen individual pictures are taken, exposing the mouth to doses as high as ten rads. If this high a dose over such a small area were extrapolated to the entire body, the levels of radiation absorbed would be enormous.

But the fact remains that dental x-rays generally do not spell so much risk as medical x-rays, because the areas exposed are very limited. If dental x-rays were the only radiation individuals received, there would be no problem. But, because other sources abound, it makes good sense to reduce such risk as is involved in dental x-rays.

Perhaps the most important consideration is necessity. It is true that the presence of certain dental problems can be confirmed only by x-ray. For example, periodic x-rays may be necessary in older patients with degenerative gum disorders to follow the progress of the disease. But is it necessary to routinely x-ray the teeth of asymptomatic children whose baby teeth will be replaced by permanent ones later on? You should ask your dentist whether your or your child's x-ray exposure, no matter how small, is justified in terms of a tangible patient benefit.

Other ways to reduce radiation risk in a dental chair include performing a visual check of the x-ray equipment. Many dentists still use units equipped with short, pointed dental cones that scatter radiation and bounce it back off the walls, thus often exposing reproductive organs to unnecessary radiation. If your dentist is still using the short, pointed cone device on his x-ray machine, ask about the new, lead-lined dental cylinder. This roughly eight-inch-long attachment helps to improve collimation (a process that limits the width of the x-ray beam), making the beam narrower than that produced by the cone-shaped tips and keeping radiation from scattering. Even with this nonscattering tip, a lead shield should be required to protect your reproductive organs.

You may also want to ask about a type of x-ray picture that produces a panoramic view. In this technique, the film is located outside the mouth, and the x-ray camera is rotated around the patient. Exposure with a panoramic x-ray is about one-fortieth that of the routine full-mouth series, and the dose delivered to the thyroid gland is roughly one-tenth as much.

Should you change dentists, there is no breach of ethics involved in your new dentist asking your former dentist for your previous dental x-rays. Suggest that the new practitioner do this to spare you the unnecessary exposure involved in a completely new set of x-rays.

CONSUMER PRODUCTS

As we noted at the beginning of this chapter, medical and dental radiation is the type that most of us are exposed to most often. Such exposure therefore offers us the greatest opportunity to reduce our risks. But what about other types of radiation generated by human activities? Are we sitting ducks for exposure from consumer and public service products? Fortunately not.

The products we refer to include static eliminators, high-voltage rectifiers and control circuits, radioactive luminescent devices, and electron microscopes—obviously a group with which very few individuals ever have any contact. But how about dental bridgework, airport luggage scanners, smoke detectors, tobacco products, and, of course, TV sets?

This class of products contributes an average yearly whole-body dose of less than 5 millirems per person. But some, including electron microscopes and tobacco products, contribute more than others. As noted in Chapter 4, cigarette smoke contains the radionuclides polonium-210 and lead-210, and the concentration of these isotopes in the lungs and bones of smokers is 30 percent higher than in nonsmokers. One of the best ways to reduce radiation exposure is simply to give up smoking.

Among products that give off small amounts of radioactivity—airport inspection systems, high-voltage vacuum switches, and certain combustible fuels, for example—perhaps the only one of any significance is the television receiver. It matters mainly because so many people spend so much time in front of it.

X-rays may be produced by electrons accelerated by the high-voltage components found in a TV set. And these x-rays can escape and irradiate the viewer. In 1969, a federal standard was established limiting x-ray emissions from TV sets to 0.5 milliroentgens per hour. All TV sets must conform to this standard.

To make sure that your new TV set is as free of x-radiation as possible, check the back for a label certifying that it conforms to federal emissions standards. This has been a requirement on all sets manufactured since 1970. Today, watching TV from the usual viewing distance should pose no hazard to health. If your set needs servicing, however, have the repair done by a qualified person to make sure that x-radiation is kept to a minimum. Trained technicians use TV x-ray detecting devices developed by the government to spot sets with higher x-ray emissions than are legal. If excessive x-ray leakage is found, ask your state or local board of health to advise you.

Finally, we come to risk reduction in the operation of nuclear reactors. As we mentioned at the beginning of this chapter, when a nuclear power plant is operating normally, there is virtually no risk to plant personnel or to residents of the area in which the plant is located. It is only when some mishap occurs that these plants put populations at risk. This was the lesson of Three Mile Island. And if there is a silver lining to that accident, it is the resulting measures designed to reduce even further the small risks involved in the operation of nuclear reactors.

For example, new research on the subject of safe reactor operation has been funded by the nuclear industry. Every reactor in the country has been linked to the government's emergency response center by means of a telephone hot line. And an intensive program of extra training for reactor personnel has begun.

More specifically, a nuclear safety analysis center (NSAC) has been set up by the industry to study the accident at Three Mile Island and relate its findings to the solution of problems at this and other nuclear power facilities. A second group, the Institute for Nuclear Power Operations, was formed to suggest ways and means of improving operating standards and ensuring that all nuclear reactors meet the standards.

Two telephone hot lines have been installed by the Nuclear Regulatory Commission between every nuclear facility in the United States and the emergency response center of the NRC in Bethesda, Maryland. One line is intended for use when emergencies develop in the operation of the reactor, and the other is for obtaining advice on radiologic emergencies. These avenues of communication ensure that the best available thinking will be brought to bear whenever and wherever it is needed.

All operators and supervisory personnel in charge of nuclear reactors have been given cram courses in appropriate procedures to use in various types of emergencies. These training sessions are carried out with the aid of special computers that can simulate just about any type of reactor accident that can occur. Furthermore, two resident inspectors assigned by the Nuclear Regulatory Commission will be stationed at every reactor in the United States. Their major responsibility will be to identify and address potential problems before they can result in a serious accident.

Admittedly more can be done, and various consumer groups are quick to point this out. Perhaps plants near densely populated areas should not be allowed to operate until the industry can prove beyond any doubt that it is able to protect the public against a serious accident.

What is encouraging is that steps can and are being taken—on an individual, industrial, and federal basis—to reduce risks inherent in the use of ionizing radiation. A tremendous force that can benefit all humanity is at stake. And just as science found a way to release the nuclear genie from his bottle, so it will find a means of minimizing the risks we run when we engage his services.

10
Reducing the Risks in Using Nonionizing Radiation

Chapter 9 suggested ways to reduce risks arising from exposure to ionizing radiation. But, aside from x-rays, ionizing radiation is not the kind most Americans are apt to encounter in the course of an average day. The same is not true for sources of *non*ionizing radiation. From the moment you step out into the morning sunshine, your exposure begins. And it probably doesn't let up until you turn your bedside light off at night. In between, you have probably been exposed to nonionizing radiation from numerous sources that you see, touch, and use constantly.

For the most part, these sources are convenience devices that you certainly wouldn't want to part with. So again it becomes a question of how to enjoy the benefits with minimal risk. Here are some ways in which you can work toward this goal.

AVOIDING OVEREXPOSURE TO ULTRAVIOLET LIGHT

Starting at the most energetic end of the electromagnetic spectrum, the first type of nonionizing radiation we encounter is ultraviolet light. And its prime source is the fountainhead of all life, our sun. Ultraviolet radiation from the sun is a leading cause of skin cancer, which causes an average of 7,000 deaths a year, especially among fair-skinned people. But unlike cosmic rays, which also arrive from outer space, solar ultraviolet rays are eminently avoidable. You can even enjoy the warmth of sunlight without any exposure to ultraviolet rays simply by letting the sun bathe you through glass. Ultraviolet rays do not have enough energy to pass through. And just by staying out of the sun, you can avoid *all* risk.

Most dermatologists do not approve of excessive sunbathing, and they think those bronzed sun worshipers you see entrenched on the beach are courting skin cancer. However, most people do look good with a tan, and you can have a tan with minimal risk if you observe one simple precaution. More than half of the ultraviolet radiation that reaches any spot on the earth each year arrives when the sun is at its zenith. The thinner the layer of earth atmosphere the rays must penetrate, the more of them get through. In the tropics, for example, the sun is almost directly overhead, and here skin burns most rapidly. If you must sunbathe, you can reduce the ultraviolet radiation and its potential biological damage by keeping to a minimum any exposure during the four-hour period between 10:00 A.M and 2:00 P.M. (but remember, this period occurs between 11:00 A.M. and 3:00 P.M. when daylight saving time is in effect). Ultraviolet radiation can also be reduced by the use of tanning creams containing materials that screen out some of the sun's harmful rays.

Some people like the tanned look so much that they try to maintain it during the winter by means of devices that produce ultraviolet light. Sunlamps, for example, can tan you beautifully, but each year almost 8,000 people sustained sunlamp-related injuries that required hospital emergency room treatment, according to the Consumer Product Safety Commission.

To reduce the incidence of sunlamp injury, in late 1979 the FDA established safety standards applicable to all instruments manufactured after May 6, 1980. These standards make mandatory certain design features that make the lamps safer to use. Included among these features are a timer that turns the lamp off automatically after 10 minutes of operation; a switch that gives the user the option of turning the lamp off manually; a special base for the sunlamp bulb, to preclude its use in a regular light socket; and enough pairs of protective goggles so that, when more than one person use the lamp at one time, all eyes can be shielded.

The standard also requires that sunlamp users be given the information they need to use these products safely. Accordingly, each sunlamp fixture must carry a warning label that bears the following statement:

DANGER—Ultraviolet radiation. Follow instructions. As with natural sunlight, overexposure can cause eye injury and sunburn; repeated exposure may cause premature aging of the skin and skin cancer. Medi-

cations and cosmetics applied to the skin may increase your sensitivity to ultraviolet light. Consult physician before using lamp if taking any medication or if you believe yourself especially sensitive to sunlight.

The label also must include information about the minimum safe distance between lamp and user, the maximum safe exposure time, a recommended tanning schedule, and a warning to use protective goggles.

The desire for a natural, year-round tan—the so-called California look—has also given rise to commercial tanning booths. These extremely popular devices measure about three to four feet square and can accommodate one person standing up. Fluorescent sunlamps are mounted in the corners, and reflecting material covers the walls. The ultraviolet rays generated in tanning booths are intense enough to cause severe "sunburn." And a user who falls into an unshielded lamp can break it and sustain serious cuts and lacerations.

Because of these problems, the FDA issued a special communication to some thirty manufacturers, urging voluntary and immediate adoption of specific safety measures. To correct possible hazards arising from use of these booths, the manufacturers were asked to install shields to prevent users from coming into contact with the lamps and physical barriers to keep them at the proper exposure distance. Also requested were more accurate timers to turn off the lamps after no more than ten minutes, protection against electric shocks and fires, physical aids such as handrails to help prevent falls, and a ventilation system to keep the temperature in the booths below 100° Fahrenheit.

Other sources of ultraviolet light include mercury vapor lamps and tungsten halide lamps, both of which can leak radiation if their outer globes of protective glass are broken. To prevent this leakage, at least two new types of mercury vapor lamps that were recently approved for sale shut off automatically when the outer glass housing breaks. Tungsten halide bulbs that operate on the same principle are being developed. But these safer lamps are still relatively new, and the older-style bulbs are still used most often.

If you live or work where mercury vapor or tungsten halide lamps are used, here are some risk-reducing steps you can take. Should you notice a broken lamp, call the local utility company or other responsible party, and say you want the lamp turned off. Then don't wait for someone to come; leave the area immediately to minimize your exposure.

If you are exposed and develop burns or eye problems, see a doctor and be sure to mention that your problems may be due to ultraviolet radiation. You should also report the incident to the FDA and your State Health Department as soon as you can.

If you illuminate your home or yard with these types of lamps, check them regularly *when they are off* for breakage or puncture—preferably before each use. Remember to replace bulbs only after they have been allowed to cool and the electricity is off. Needless to say, if your lamps are not of the type that shuts off automatically when the globe is punctured, you should consider replacing them with the newer types that incorporate important new safety features.

GUARDING AGAINST HARMFUL LASER BEAMS

Continuing toward the less energetic end of the electromagnetic spectrum, we come next to the range of visible light, which includes lasers. You will recall from Chapter 7 that very few states have enacted laws designed to reduce risk from the laser beams that one might encounter at, say, a light show.

For your own protection, remember that total radiation exposure in the audience area, including radiation reflected from mirrors and other scattering devices, cannot by law be greater than that which has been found to cause biological damage. This is the Class I laser category, and the effects produced by such devices are not very spectacular. So, if you are being "awed and amazed" by a light show, chances are that the laser beam is much stronger than the legal limit. In such cases, make sure that the path of the beam is well over your head. FDA guidelines require that the path of the beam be about 20 feet (6 meters) above the audience if the device is one that runs without an operator and half that distance if an operator is there at all times.

Further, even when the laser device can run without an operator, the FDA requires that one individual be responsible for shutting down the machine if problems develop. You may want to check with local or state officials to be sure that all their requirements have been met and clearances obtained before you or any members of your family attend a laser light show.

PROTECTING YOURSELF FROM INFRARED LIGHT

Infrared radiation can produce adverse biological effects on tissue through the action of heat. These effects can range anywhere from a temporary reduction of sperm count, when heat is applied to the testicles, to temporarily arrested breathing in infants exposed to radiant warmers. For this reason, the safety of infrared lamps used to warm the air in operating rooms where infants undergo surgery is under investigation. You may also be interested in bringing pressure to bear on state and federal legislators to support the establishment of standards for exposure of the skin or eyes to infrared sources.

Other risks involved in infrared radiation can be reduced through certain specific measures. If you work with or use any infrared source, make sure that its surfaces are well insulated to reduce the heat given off by the device. Aluminum is often used for this purpose because of its ability to reflect heat, and industrial workers can reduce their risk by wearing aluminum aprons, coats, or gloves. All skin surfaces should be protected by some type of clothing; lightweight materials are most desirable for this purpose.

This area is ripe for more research. Before definitive infrared exposure standards can be set, scientists need to know much more about threshold limits, the effects of time, and the biological consequences of low-intensity exposure over prolonged periods of time. You can help by urging state and federal legislators to support this research.

PRECAUTIONS TO HEED IN USING MICROWAVE OVENS

Proceeding through the electromagnetic spectrum toward the less energetic end, we come next to the microwave frequencies. Microwaves enjoy wide use in a variety of radar systems. Their ability to bounce off metal and return to a sensing device in an absolutely predictable manner makes these waves particularly useful for police detection of speeding cars, early detection of possible attack by enemy aircraft or missiles, air traffic control, satellite communication, and transmission of telephone and television signals.

The risk involved in the use of radar devices is so slight that the benefits clearly outweigh it. In fact, talking about reducing the "riskiness" of radar devices is nothing more than an academic exercise that needn't detain us here. However, the broadest consumer use of microwave energy is in the so-called space-age, or microwave, oven. Without a doubt, this most wanted of kitchen conveniences is still growing in popularity. And with this device, discussion of how to reduce the risk *is* appropriate.

Since 1971, all microwave ovens manufactured in the United States have been regulated by the FDA. And this agency believes that ovens meeting its standards and used as the manufacturers suggest are completely safe. In practice, what does this mean?

The FDA radiation safety standard limits the level of microwave leakage from an oven *throughout its lifetime* to 5 milliwatts of microwave radiation per square centimeter at approximately 2 inches from the oven surface—a level well below that at which biological damage is known to occur. And no radiation is left over when the machine is turned off, just as there is no afterglow when an electric light is extinguished.

The FDA also took further steps to ensure the safety of microwave ovens. All such products manufactured after October of 1975 must carry a label explaining precautions for safe use. This label must clearly display the following statement:

> *Precautions for safe use to avoid possible exposure to excessive microwave energy. Do not attempt to operate this oven with: (a) object caught in door, (b) door that does not close properly, (c) damaged door, hinge, latch, or sealing surface.*

This label must be clearly visible, even when the oven is in use.

But even if FDA standards are vigorously met, a microwave oven can be dangerous if it isn't used or serviced properly. You will probably be able to identify such obvious problems as damaged hinges, latches, or seals. But if these defects are not present and you are still concerned about excessive leakage, call the manufacturer of your oven. Many provide the service involved in checking ovens for leakage. If your unit was made by a manufacturer who doesn't, contact a service organization that specializes in microwave ovens. Many state health departments maintain funded programs for oven inspections in your home. At the very least, they will give you the names of reputable service organizations with the proper equipment to test your oven for excessive emission.

Finally, you may want to call a regional office of the FDA. This agency tests a limited number of microwave ovens in homes as part of an overall program to assure compliance with federal requirements. The FDA further conducts a program for spot-checking microwave ovens on commercial premises, at the establishments of dealers and distributors, and in its own laboratories, so that all possible hazards can be eliminated. It also assesses the radiation testing and quality control programs run by manufacturers. If a defect is found, the FDA requires the manufacturer to make appropriate repairs that bring the unit into compliance with federal safety standards—at no charge to the owner.

You should *never* try to test for microwave leakage yourself. In recent years, a microwave testing device came on the market for sale to consumers. But when a representative sample of these devices was tested by the FDA, they were found to be generally inaccurate and not reliable for anything but very approximate readings. You can depend on the testing instruments used by public health officials, because these devices themselves are regularly tested for accuracy.

Of course, the availability of testing procedures does not relieve the microwave oven owner of the responsibility to operate the unit correctly. Here are a few general guidelines to help you operate your microwave oven safely.

Make sure the unit has not been damaged in transit. When your new oven arrives, examine it and the shipping carton carefully for any evidence that it was dropped or otherwise handled roughly. This kind of treatment can cause damage that results in microwave leakage.

Follow directions. The manufacturer always supplies an instruction manual with the unit. Adhere strictly to all recommended methods of operation, and heed all safety precautions.

Never use the oven if the door doesn't close completely. Be on the lookout for a bent or warped door, especially if the door has been jostled heavily while open. Even slight misalignments can release radiation.

Always insert objects correctly. You should never try to put anything into the oven through the door grill or around the door seal.

If the oven is on, make sure there is something inside. Heat energy must be dissipated somehow. If there is nothing to absorb it, the microwave energy can damage the unit.

Make sure the door and seals are clean. Use water and a mild detergent (never scouring pads, abrasive cleansers, or steel wool) to remove grease that can accumulate around the door seal and allow radiation to escape. Special microwave oven cleaners can be used, but they are not necessary.

Never tamper with oven safety interlocks. They are there for your and your family's protection. And allow nothing, not even a paper towel, to stick through the door crack.

Make sure your unit is serviced regularly. A microwave technician should check your oven periodically for damage, wear, tampering, or radiation emission. And if you notice any deterioration of seals around the door in the interim, have them replaced immediately by a qualified person.

Don't take your microwave oven abroad. Most foreign countries do not use the 110/220 current found in the United States. If you bought your unit in this country, it will not work elsewhere. Nor can it be safely rewired or otherwise adapted for a different current.

Do not use your oven for home canning. The FDA does not believe that microwave units produce high enough temperatures to kill the disease-causing bacteria that are often found in foods being canned.

If your oven was manufactured before 1971, keep it at arm's length. Units made before this date may not conform to FDA safety standards. Just to be on the safe side, keep a good few feet away, except to turn the oven on or off. Remember, distance is your greatest protection. Even if your oven does leak, at twenty inches away you will receive only about one-hundredth of the radiation you would absorb at a distance of two inches.

By following these guidelines, you will be able to take advantage of the convenience microwave ovens offer, while reducing any risk their use may entail.

We have tried in this chapter to suggest ways in which devices that emit ultraviolet radiation, visible light, infrared radiation, or microwave radiation can be enjoyed for the benefits they offer—with minimum risk. Chapter 11 addresses protective measures that do not involve actual contact with that not-so-humble servant, radiation.

11
Safety Innovations

One of the most encouraging developments in the minimization of radiation risk is an increasing dialogue on the subject among members of the medical community. And the doctors are speaking not so much of specific measures as of general ones that take into account all aspects of the problem. This orientation is important, because it encourages every professional concerned with medical radiation exposure to view his or her involvement not as a function in and of itself, but as one link in a chain.

Health insurance plans, for example, which may insist on pretreatment dental x-rays for many patients to substantiate that the work done by a dentist was really needed, are now looking at the problem of unnecessary radiation exposure from a broader viewpoint. To cite another example, certain hospital emergency room procedures are being re-evaluated. Often, a person coming to an emergency room is first interviewed and screened by a secretary or nurse, and x-rays may be ordered even before a physician has seen the patient. These x-rays may be unnecessary, they increase the patient's overall exposure to ionizing radiation, and they add to the cost of the visit unnecessarily.

As you pursue your own efforts to avoid unnecessary radiation exposure, it may be useful for you to know about some of the safety innovations being discussed today. Your doctor and dentist are almost certainly as concerned as you are about your radiologic health. They would probably be happy to talk with you about it, and out of such discussion can come constructive ideas based on *your* particular circumstances. This is really a key to minimizing exposure risk: the willingess of professionals to think of their particular functions more broadly and of their patients in more specific terms.

REDUCING THE NUMBER OF X-RAY EXPOSURES

One way to make diagnostic x-rays safer is to minimize the *number of x-rays* patients submit to in their lifetimes and, in the case of pregnant women, to avoid risking damage to the unborn. Several "procedural" innovations can help physicians and their patients reach this goal.

The only time you should be given an x-ray examination is when you need it. In other words, a diagnostic x-ray examination must be tailored to your particular profile and take your "specifics" into account—your age, type of illness or injury, symptoms, family medical history, and radiation history, for example.

The National Council on Radiation Protection and Measurements provides some guidelines for the radiologist who will take the x-rays, and for the physician who orders them, for a woman of child-bearing age. Included in this list are three basic directives:

1. Any time a nuclear examination or x-ray of the pelvic or abdominal area is ordered for a woman in her child-bearing years, a note should be attached to the order form stating that the possibility of the patient's being pregnant has been considered.

2. When a request for x-rays is made on behalf of a patient known to be pregnant, her doctor should discuss with the radiologist how best to conduct the examination, so that useful information is conveyed to the doctors but risk to the fetus and mother is limited.

3. When the woman is definitely not pregnant, she should be advised that becoming pregnant might entail a risk. An ovum damaged but not killed during the examination may be subsequently fertilized, and the effects of these events on the fetus are just not known. Thus it is advisable to wait until two months after the examination of her pelvis or abdomen before attempting to become pregnant. This waiting period tends to obviate any problems.

Before deciding to take x-rays, a doctor should at least take a look at what has been done before. The problem is that most doctors do not have time to contact their patients' former physicians and obtain existing x-rays. One possible solution is to make *patients* the official custodians of the films taken whenever they visit a medical or dental facility.

The NCRP recommends that the maximum permissible dose to the fetus resulting from *occupational* exposure of the mother not exceed 500 millirems. This creates a certain kind of job discrimination. The implication is that, even if they are not pregnant, fertile women should be employed only in jobs where the annual radiation dose is lower than that recommended for other occupational radiation workers. Thus some jobs would be closed to large numbers of workers strictly on the basis of their sex and age.

Another unique feature of this recommendation is that it treats the unborn child as an already existing member of the general public, who has been brought involuntarily into a workplace where exposure to radiation is possible. Thus the unborn child's sensitivity to leukemia or other disease or abnormality dictates where the mother may work, even though she runs no risk herself.

If this idea were to be implemented, differences in radiologic technique and x-ray quality would certainly have to be taken into consideration. It would be absolutely necessary to label films carefully to make them as useful as possible to future consultants. Without such information as the patient's name, date of the x-ray, and orientation of the patient, repeat x-rays would have to be made.

Legally, there is no clear-cut answer to the question of whether the doctor or the patient owns x-ray films. More and more doctors, mindful of their responsibility not to contribute to increased x-ray exposure of their patients, are letting the patient have the films. You may therefore want to ask for any pertinent x-rays before you see a new doctor or dentist. If your doctor does not want to release the original x-rays, try suggesting that copies be made for you.

Another point you may want to bring up with your physician is the practice of destroying x-rays after five to six years on the grounds that the patient's condition has probably changed so much that the films have no further use. This is not necessarily true. Very often, such x-rays help physicians considerably in diagnosing and treating certain diseases. "Old" x-rays can be placed on microfilm and given to patients by hospitals and physicians, so that an individual can keep a record of his or her own x-ray history for future reference by other physicians.

The promotion of what are known as referral criteria has been a positive development in reducing radiation exposure. These criteria were developed by experts from various government bodies convened by a 1978 directive from the White House to the secretary of health, education and welfare. Dissemination of the criteria is conducted under the auspices of the Bureau of Radiologic Health. The term referral criteria simply refers to specific information—signs, symptoms, diagnostic possibilities based on previous examinations and family history, and, of course, diagnosis already established—that guides health practitioners in deciding when to send their patients for an x-ray examination. In addition to purely medical considerations, specific criteria should be available to guide physicians through the maze of their legal responsibilities in ordering or not ordering different radiological examinations. (In many hospital emergency rooms, attending nurses can decide on their own that x-rays should be taken, even before the patient has been examined by a doctor. That nurses should be specially trained to make these decisions goes without saying.)

COMPUTER CONTRIBUTIONS

Computer technology offers another interesting means of reducing exposure to radiation. For example, computers can make available complete information on the various radiation sources present around us and specific data on levels of radiation exposure to the general public. On a more practical level, computers can be applied to the reduction of x-ray exposure among individual U.S. citizens. This goal could be achieved by issuing, at every patient's *first* x-ray examination, a computerized plastic card of the type that banks issue as charge cards. These cards are designed to be inserted into computer terminals at bank branches and provide readouts of the account balance. The x-ray card would bear the patient's name and social security number or other identifying number. All the patient's personal x-ray history would be coded and maintained under that number.

Before any medical or dental x-ray, nuclear medicine, or fluoroscopy procedure could be performed, the plastic card would be inserted into a computer terminal by a health care professional, who would then receive a readout of the patient's total, lifelong x-ray history. If the total number of rems that patient had received was on the high side, or if the patient had recently undergone a similar x-ray or nuclear scan at another health-care facility, the physician might consider postponing, or even canceling, the procedure. This system would be be especially useful for patients who are unable or unwilling to maintain their own record of their x-ray history.

Another possible use of computers is suggested by the computerized matching of precoded information that is presently being done with bank cash machines and airport traveler's check machines. A card is inserted into a terminal and assets are dispensed only if the right combination of numbers is fed into the machine. Using this technology, x-ray machines could be designed not to operate unless all conditions necessary for minimal x-ray dose (proper screens, proper type of film, distance from the patient, and so on) were met.

TRAINING, LICENSING, AND EDUCATION

The most common cause of x-ray overexposure is technical mistakes made by inexperienced or poorly trained technologists. Yet only a handful of states now require x-ray technicians to be licensed or registered. Concerned citizens should work with their elected legislators to establish training and registration programs.

For example, technologists should be keenly aware that the size of the x-ray beam must be limited and that it must be aimed carefully at only those areas of the body that are to be examined. They should also be aware of the importance of doing things right the first time, so that repeat exposures will not be required. The number of films taken is drastically reduced when technologists simply make careful measurements by means of calipers, use phototiming devices on the x-ray machine, and refer to exposure charts to make sure that the camera is set right the first time.

Even physicians not specifically trained in radiology may not be sensitive to the biological effects of radiation or knowledgeable about protecting patients from it. Perhaps the federal or state government should license the sale of such equipment only to individuals who can demonstrate that they have met rigorous standards of expertise in the use of radiation.

It would also make good sense to educate all doctors in the efficient use of radiation. It is time for all medical schools to include in their curriculum courses in radiation physics and safety. A similar need exists for courses designed for dentists and dental technicians. Dental schools should appoint dental radiation officers and include courses on dental radiation in their curricula.

In the general area of better education about radiation, the Bureau of Radiologic Health publishes a series of communications that often contain suggestions for reducing exposure to radiation. If your doctor or dentist is not aware that this information is available, by all means mention it. These BRH communications should be more widely circulated, so that not only radiologists, but other physicians and dentists using x-ray equipment as well, can benefit from the suggestions provided.

Patients themselves should know more about the risks and benefits of medical procedures that involve ionizing radiation. (This book, of course, was conceived to fill that need. But this subject should also be covered in school, perhaps as a part of hygiene or biology courses in junior high school. After all, the science of radiation is going to play a major part in the lives of all Americans in the years to come, and every citizen should be able to make educated decisions.

REDUCING THE LEVEL OF EXPOSURE PER X-RAY

Another sound way to reduce lifetime exposure to radiation is to minimize the *exposure required per x-ray*. A lively topic in the medical press today is the sacrifice of some clarity in x-ray films for the purpose of reducing patient exposure. There are times when exquisite radiographic detail is not essential—for example, in barium enema examinations, gall bladder studies, or examination of the spine. Furthermore, certain techniques can reduce radiation exposure to the patient while only minimally reducing the clarity of the picture—for example, the use of intensifying screens that employ chemical compounds referred to as "rare earths."

The quality of an x-ray picture depends on how much radiation is stopped by the intensifying screen and on how well its energy is converted into light. Rays that pass through the screen are not converted into light at all. Rare earth screens are as much as 2.5 times as efficient as other screens in blocking x-rays. Thus they allow the speed at which the picture is taken to be doubled and the time of patient exposure to be just about cut in half.

There are other means by which patient exposure can be reduced. Incorporating graphite (which absorbs little radiation) into the top of the x-ray table and the x-ray cassette allows more radiation to enter the cassette and reach the film, producing a better image with a lower rate of exposure to the patient. And the use of special high-speed film can considerably reduce radiation exposure with very little loss of detail. Even the actual number of films taken during a given examination can often be reduced. Certain views can be eliminated in tomography, angiography (x-rays of blood vessels), barium enemas, upper G.I. series, mammography, and spinal films, among others.

"QUALITY ASSURANCE"

Quality assurance (QA), according to the Bureau of Radiologic Health, is the "planned and systematic actions that provide adequate confidence that a diagnostic x-ray facility will produce consistently high-quality images with minimum exposure to patients and healing arts personnel. The determination of what constitutes high quality will be made by the facility producing the images."

Behind the convoluted officialese is a fairly simple idea: a concerted effort to improve x-ray quality and reduce radiation exposure. QA is voluntary, and all professionals who use x-rays (radiologists and other physicians, as well as chiropractors, podiatrists, and dentists) are being encouraged to establish their own QA programs.

Innovations may include changing established behavior and work patterns, purchasing equipment that conforms to official performance standards, monitoring the quality of images as well as the functioning of equipment, and keeping records of how often and why x-rays are retaken.

Formal QA programs structured along these lines are now being actively promoted by three major segments of the radiologic field: professional radiologist organizations, the federal government, and several manufacturers of x-ray equipment and film. Yet despite their efforts, interest in QA programs is not particularly high. Perhaps the worthy objectives of QA programs could be achieved if pressure were brought to bear by the general public.

Safety innovations such as we have discussed in this chapter are dear to the hearts of radiologists. All too often, critics of radiology pillory the profession as an isolated group of zealots. Nevertheless, it is unfortunate that no means exists for determining whether a diagnostic procedure has been of clinical value in each and every examination. Nor is there an impartial method for determining whether a particular x-ray examination is really necessary.

Physicians practice in a world of probabilities. The outcome of their efforts on behalf of a given patient is unpredictable. In their quest for an early diagnosis of disease, they may sometimes do more harm than good. But to practice medicine any other way would be far more harmful to more patients in the long run. Attempts to legislate matters of medical judgment can only be self-defeating, and it would be wrong to restrict the use of radiation further and further without definitive data to support such action.

part four

The Role
of Government

12
The Nuclear Regulatory Establishment

As an individual concerned with the use and possible abuse of radiation, you naturally want to stay abreast of new developments. The mass media unfortunately don't always supply all the information we need to form sound judgments. Too often, newspapers feature stories calculated to increase circulation rather than presenting unbiased, objective reports. You may therefore want to obtain your information from official sources. You may also have questions about various aspects of radiation sources that impinge on your life, or you may wish to call the attention of an official government bureau to some misuse or possible radiation hazard.

We conclude this book with an outline of the nuclear regulatory establishment—the various government agencies involved in radiation protection—including the nature of their involvement and their areas of responsibility. Do not hesitate to use their services. These organizations are funded by your tax dollars and exist solely to advance the interests of U.S. citizens.

DEPARTMENT OF HEALTH AND HUMAN SERVICES

Heading the list is the vast complex that until mid-1979 was known as the Department of Health, Education and Welfare and is now called the Department of Health and Human Services. Three bodies in the Public Health Service of this department are of interest to us: the Food and Drug Administration (FDA), the National Institutes of Health (NIH), and the Center for Disease Control.

UNITED STATES GOVERNMENT RADIATION PROTECTION
RESPONSIBILITIES (As of January 1, 1980)

DEPARTMENT OF HEALTH AND HUMAN SERVICES	
Food and Drug Administration	Radiation emissions from medical devices and consumer products, radioactivity in foods and drugs, research
National Institute of Health (National Cancer Institute, National Institute of Environmental Health Sciences)	Radiation health effects research
Center for Disease Control National Institute for Occupational Safety and Health	Research on occupational exposure, criteria documentation for OSHA
DEPARTMENT OF LABOR	
Occupational Safety and Health Administration	Health and safety standards development and enforcement for radiation workers
Mine Safety and Health Administration	All mining radiation exposure
DEPARTMENT OF ENERGY	Weapons development, fuel fabrication and enrichment, reactors, accelerators, research
DEPARTMENT OF DEFENSE	Naval propulsion, weapons deployment and maintenance, radiation safety for military personnel, research
DEPARTMENT OF TRANSPORTATION	
Federal Aviation Administration Federal Railroad Administration Federal Highway Administration	Exposure from transport of radioactive materials— workers, passengers, and general public

DEPARTMENT OF COMMERCE	Coordination of nonionizing radiation research
National Bureau of Standards	National standards for measurement and quantification of radiation, physical research, calibration services
ENVIRONMENTAL PROTECTION AGENCY	Radioactivity in air and water, Federal Radiation Council authority
NUCLEAR REGULATORY COMMISSION	Licensing of nuclear power facilities; regulation of by-product, source, and special nuclear material
CONSUMER PRODUCT SAFETY COMMISSION	Consumer products that use radioactive materials other than those regulated by MRC
FEDERAL COMMUNICATIONS COMMISSION	Radiation exposure from satellite and communications systems
VETERANS ADMINISTRATION	Radiation safety for medical radiology, nuclear medicine, and research by VA employees

Adapted from material supplied by the U.S. Department of Health and Human Services, Public Health Service, Food and Drug Administration.

FOOD AND DRUG ADMINISTRATION

This organization is made up of eight bureaus involved with foods, drugs, veterinary medicine, biologics, medical devices, research on poisons, regional operations, and radiologic health. Because of the scope of this book, we will limit our discussion to the Bureau of Radiologic Health (BRH). The overall functions of this bureau include development of programs designed to control possibly hazardous exposure to ionizing and nonionizing radiation and to assure its safe use. Toward this end, the BRH conducts electronic radiation-control programs and sets performance standards. It supervises compliance with radiation exposure limits, plans and oversees research on radiation exposure, develops programs for reducing radiation exposure, and shares its expertise with other agencies involved in radiation.

One arm of the BRH is divided among four separate divisions. The *Division of Biologic Effects* coordinates research on the effects of radiation. It includes the Experimental Studies Branch, composed of sections devoted to a wide range of scientific disciplines, the Epidemiological Studies Branch, and a Standards Support Staff that supervises standards and guidelines.

The second important arm of the BRH is the *Division of Compliance,* the primary function of which is development of criteria, standards, and regulations that protect the public from injurious radiation emitted by various products and materials. These duties are carried out by a Consumer-Industrial Products Branch, which formulates compliance standards for such products as TV sets, light or laser devices, and microwave products; an X-Ray Products Branch, which deals exclusively with standards for this class of products; a Standards and Regulations Branch, which coordinates the development of all regulations under authority of the BRH and maintains official public files for all appropriate documents; a Compliance Data Branch, which develops and maintains all systems of compliance information; and a Compliance Operations Branch, which prepares and implements all compliance programs.

Third of the four BRH components is the *Division of Electronic Products,* which studies and evaluates conditions of exposure and emissions of radioactivity from various products. Separate types of products are handled by the Electromagnetics Branch (radio and microwave frequency products); the Electro-Optics Branch (infrared, visible, and ultraviolet light); the Radiation Physics and Electronics Branch (products that cause ionizing radiation due to alpha particles, beta particles, or gamma rays); the Acoustics Branch (concerned primarily with sound energy); the Program Support Branch (administrative, design, and construction services); the Medical Physics Branch (ionizing radiation products used in the practice of medicine); and the Nuclear Medicine Laboratory (radiopharmaceuticals).

Rounding out the divisions of the BRH is the *Division of Training and Medical Applications.* It is involved in a nationwide program to reduce unnecessary exposure to radiation in the healing arts. This division is supported by an Analysis and Evaluation Branch that conducts field investigations and spots trends; a Medical Branch that concentrates on medical diagnostic radiation; a Quality Assurance Branch that addresses quality and image-exposure problems in x-ray equipment; a Program Support Branch dealing with budget and management matters; a Radiation Therapy Branch that works to improve patient care in both ionizing and nonionizing radiation therapy; and a Nuclear Medicine Branch charged with the responsibility of promoting patient safety in nuclear medicine.

The BRH also promotes consumer and patient education about products that emit ionizing or nonionizing radiation. Among the materials they provide to advise the public on how to minimize radiation risk are slide/tape programs; posters; and an x-ray record card of the type mentioned in Chapter 9, on which patients can keep track of their x-rays and where the films are filed. The FDA also broadcasts regular public service announcements on the radio and publishes a monthly magazine, the *FDA Consumer*.

NATIONAL INSTITUTES OF HEALTH

Not a part of the FDA, but still under the aegis of the Public Health Service of the Department of Health and Human Services is the agency known as the National Institutes of Health (NIH). A major offshoot of this organization is the *National Cancer Institute* (NCI). This institute conducts cancer research and publishes a journal devoted to cancer research. It also develops less structured channels of communication, such as symposia and workshops, through which scientists can exchange cancer research information.

CENTER FOR DISEASE CONTROL

On a comparable level with the NIH is the Center for Disease Control and its subsidiary organization, the National Institute for Occupational Safety and Health (NIOSH). This organization conducts research and makes recommendations dealing with the elimination of radiation hazards in nuclear power plants, factories, and other places of employment. NIOSH provides technical information on request, and its publications are disseminated on request to occupational safety and health professionals and to the general public.

DEPARTMENT OF LABOR

At the Labor Department, NIOSH recommendations are received by the Occupational Safety and Health Administration (OSHA), which uses them to develop health and safety standards for radiation workers, as well as methods of enforcement. Also within the Department of Labor, the Mine Safety and Health Administration regulates safety and health standards for mines.

DEPARTMENT OF ENERGY

Running down the roster of government bodies concerned with radiation protection, we come to the Department of Energy (DOE). Charged with the responsibility to oversee weapons development, fuel fabrication and enrichment, reactor and accelerator operation, and research, the DOE includes the offices of Public Affairs (all contacts with the news media), Congressional Affairs (departmental contacts with the legislative branch of government), Consumer Affairs (liaison with consumer protection agencies), Intergovernmental Affairs (working with other government agencies as conditions dictate), and *Education, Labor and Business Affairs* (contacts with these three components of the private sector).

The DOE also maintains offices to regulate the use of energy: the Economic Regulatory Administration, which keeps an eye on energy costs, and the Federal Energy Regulatory Commission, a body that supervises energy use within the government. Information on departmental activities can be obtained from public information offices at DOE regional branches and from government-owned national laboratories operated by contractors. Through the DOE Office of Health and Environmental Research (OHER), training programs on radiation protection and the handling of nuclear accidents are conducted by the Oak Ridge Associated Universities. A number of films on radiation are available to the public through the Public Affairs Office, as are various printed materials.

DEPARTMENT OF DEFENSE

The Department of Defense (DOD) is responsible for naval propulsion, weapons deployment and maintenance, research into military uses of radiation, and radiation safety for military personnel. The DOD provides information to four basic groups—occupational personnel who work with weapons, reactors, or propulsion; bystanders near radiation sources, such as security guards and hospital personnel; DOD "outsiders" who, while rarely exposed, are still kept informed on radiation matters; and the public. Because of limitations on its spending, the DOD cannot conduct any formal public education campaigns. It does, however, answer questions from private individuals or the news media.

DEPARTMENT OF TRANSPORTATION

Unlikely as it may seem, the Department of Transportation also gets into the radiation protection act. Because radioactive materials and goods must be transported, this department assumes a measure of responsibility for workers, passengers, and the general public. Among the DOT organizations involved are the Federal Aviation Administration, the Federal Railroad Administration, and the Federal Highway Administration. Authority among these groups is now coordinated by the Materials Transportation Bureau. The DOT conducts no formal educational campaign on radiation hazards involved in the transport of radioactive materials, but a question addressed to the appropriate administration usually finds its way to someone who can provide the answer.

DEPARTMENT OF COMMERCE

Among the broad and varied responsibilities of the Department of Commerce are coordination of research and investigation into nonionizing radiation. An operating arm of this Department, the National Bureau of Standards, sets standards for both detecting the presence and measuring the amounts of radiation, provides services for calibrating scientific instruments, and conducts a certain amount of research. The Office of Radiation Programs compiles information on the radiologic quality of the environment in the United States and periodically issues reports on the subject.

ENVIRONMENTAL PROTECTION AGENCY

Several agencies outside the major departments of the government are also responsible for reducing the risk of radiation exposure to the public. The Environmental Protection Agency (EPA) is empowered to monitor our air, rivers, and streams for evidence of radioactive pollution and to address the long-term problem of radioactive waste disposal in deep ocean sites.

Public educational activities for this department are carried out by two subsidiary organizations—the Office of Public Awareness and (to a much lesser degree) the Office of Radiation Programs.

The Office of Public Awareness provides the general public with information on all EPA programs, including radiation programs. It develops and distributes educational materials, lends support to legitimate citizen participation in efforts to maintain a clean environment, and monitors public attitudes and news developments. Among the informational materials produced by this office for national distribution are research reports, pamphlets and brochures, audiovisual presentations, and exhibits. It also works with citizen groups interested in or affected by programs or policies of the EPA.

The Office of Radiation Programs limits its public-information activity to direct communication with people or groups who are concerned with a particular aspect of radiation in the environment. The office makes a number of reports (largely technical) available to individuals and organizations. Of more general interest are two reports published by this office. One is on the radiation protection activities of all government agencies, and the other is on the radiologic quality of the environment.

NUCLEAR REGULATORY COMMISSION

Next on the list of concerned government agencies is the Nuclear Regulatory Commission (NRC), which was formerly called the Atomic Energy Commission. This body is responsible for licensing all nuclear power facilities and for regulating all sources of radioactivity, by-products, and special nuclear material. Organizational units include the offices of Nuclear Reactor Regulation, Nuclear Material Safety and Safeguards, and Nuclear Regulatory Research. Two additional offices, Inspection and Enforcement and Standards Development, provide support for the Office of Public Affairs.

The NRC issues public announcements and news releases daily to the news media. And every week it sends compilations of these releases to thousands of information sources in the industrial and scientific communities and in the public sector.

Information on the status of licensed nuclear facilities or those under construction is also available, as are reports of any unusual occurrences at operational nuclear plants. These are supplied to the public through the National Technical Information Service, Springfield, Virginia, or the U.S. Government Printing Office. All NRC documents are listed in various professional listings and in a monthly list published by the NRC.

To supply data consistent with the public's right to know, the NRC maintains a comprehensive Public Document Room at 1717 H Street N.W. in Washington, D.C. Almost a quarter of a million documents comprising roughly 5 million pages are available to the 7,000 individuals and groups who take advantage of this facility each year, including about 4,000 people who visit the NRC Public Document Room personally.

Detailed data on specific nuclear plants currently operating or under licensing review can be found in public document rooms at more than 130 locations around the country—usually in libraries near nuclear plant sites.

CONSUMER PRODUCT SAFETY COMMISSION

In addition to the FDA and the NRC, the Consumer Product Safety Commission (CPSC) is also charged with responsibility for the control of consumer products that may emit radiation. The FDA's authority, of course, exists basically in the radiation-emitting product area, and the NRC exercises control, under the Atomic Energy Act, over radioactive materials contained in consumer products. The CPSC, however, has similar authority to regulate the manufacture and marketing of consumer products that contain either natural or other radioactive materials. The division of responsibility was much more clear-cut before the advent of consumer products that emit radiation, but now a certain amount of overlap is unavoidable. Here

again, no formal public information services are available, but a letter
to the commission will usually be answered or referred to an appropriate source.

OTHER FEDERAL AGENCIES

Somewhat different, but related, responsibilities are assumed by other
federal agencies. These include the Federal Communications Commission (FCC), because of its involvement with radiation exposure
from satellite and communications systems, and the Veterans Administration (VA), which is responsible for the safety of VA employees involved in medical radiology, nuclear medicine, and
radiation research.

Toward this end, the VA maintains close contact with such organizations as the Department of Health and Human Resources, the
Department of Defense, the Center for Disease Control, and the
Environmental Protection Agency. It regularly consults the American Medical Association to coordinate its various programs with
medical care delivery. The VA's *Annual Report* is made available to
all news media, other government agencies, and public groups, and
it publishes a *Fact Sheet* designed to help veterans make claims for
service-connected disability. If you or someone you know may have
become disabled as a result of exposure to radioactivity while in the
service, contacting a regional VA office would be entirely appropriate.

On more general terms, in 1978 (the last year for which statistics
are available), the U.S. government spent more than $75 million for
research on the biological effects of radiation. Included were follow-
ups on the effects of radiation on the health of human populations
exposed to radiation in the past, such as the Hiroshima and Nagasaki
victims; experimental studies of radiation effects on living tissue,
using laboratory animals; and research designed to determine how
radioactive materials distributed in the environment reach human
beings.

These programs were funded by the Department of Health and
Human Services and distributed by the National Cancer Institute,
the National Institutes of Health, the Bureau of Radiologic Health,
and the National Institute of Occupational Safety and Health, within
the Center for Disease Control.

STATE AGENCIES

Like the federal government, the states of the union actively involve themselves in the regulation of radiation and in the protection of their residents from radiation hazards. All 50 states, the District of Columbia, and the Commonwealth of Puerto Rico now carry laws on their books that address the regulation of ionizing radiation.

State activities usually supplement federal efforts in radiation control. But recently, many states have begun to take upon themselves regulatory responsibility in areas that fall under the authority of various federal agencies. Thus a number of state governments (and even some municipal governments) have issued regulations restricting the transport of radioactive materials over their roads. Others are also making determined efforts to solve the problem of how to dispose of radioactive waste material.

An area traditionally left to the states and professional societies is control over the medical use of radiation. Compared to the extensive apparatus that exists for federal regulation of nonmedical use, there is practically none in the healing arts. Why? The regulatory process would be extremely time consuming and the cost prohibitive. Furthermore, the United States does not presently subscribe to socialized medicine, and there has been little inclination to impose federal bureaucratic controls on a doctor's judgment. State law, rather than federal law, governs medical malpractice liability. Because regulation involving so much judgment is an extremely thorny issue, most states delegate the responsibility for patient safety to the medical profession itself.

Thus, while the BRH does not hesitate to set performance standards for radiation-emitting products such as x-ray machines and accelerators, it is without authority to regulate *how* such devices are used. Similarly, the Nuclear Regulatory Commission avoids becoming enmeshed in the ticklish area of medical judgment about the use of radiopharmaceuticals, even though it is within the NRC's province to regulate such materials. So zealously is the principle of nonintrusion observed that the NRC recently amended its licensing regulations on the use of radioisotopes to permit physicians greater freedom and latitude in working with diagnostic radiopharmaceuticals.

In the U.S. Congress, state and national interests merge. Congress has recently become actively interested in what is being done to promote radiation health and safety. In fact, the Senate Committee on Commerce, Science and Transportation has held hearings aimed at pinpointing specific gaps in scientific knowledge related to radiation exposure. This committee, incidentally, would also like to eliminate much of the duplication and overlapping of current federal regulations. Among the committee's prime interests are framing recommendations for approving devices that emit radiation, developing yardsticks of efficacy for radiation procedures and how they are performed, and fostering a much clearer public understanding of how the health benefits from radiation-related products and services compare with the risks of exposure.

WHO SHALL SET THE STANDARDS?

Who should set standards and make determinations of which radiation exposures, on balance, benefit the general population and which do not? Representatives of private industry would leave such decisions to the government in the belief that nonprofessional individuals have neither the desire nor the background to weigh risks versus benefits.

Many government people believe that determining what constitutes acceptable risk is the province of political bodies such as the U.S. Congress and the various federal regulatory agencies. These organizations would act on the advice of affected, interested, and qualified individuals, with decisions subject to review by the courts.

However, other groups do not trust the government to make any decisions that are free of political or financial motivation. Some environmentalists would like to leave judgments about the acceptability of risk to the workers and people living near nuclear plants who must run these risks. But labor unions accuse industry of neglect in educating their workers about the potential risks involved. And so it goes, round and round.

One idea with some merit has the government establishing and maintaining a registry of radiation workers. In these files would be entered an employment history for each radiation worker over the course of her or his occupational life and a record of the cumulative dose received. Thus all individuals could be carefully tracked (in terms of occupational exposure, at least), regardless of whether they changed jobs or not.

In discussions about government involvement with radiation and the setting of safety standards, one principle seems to enjoy universal acceptance. The idea is embodied in the acronym ALARA—*a*s *l*ow *a*s *r*easonably *a*chievable. This increasingly popular term, and what it stands for, can very well become a touchstone for anyone concerned about radiation exposure.

Epilogue

We have come to the end of the book, but the story is far from finished. Subjects like ancient history and Latin poetry are complete and remain forever fixed in time. Not so the study of radiation. Even as we researched and wrote, the topic continued to evolve, with new developments arising on an almost daily basis. Capturing our topic intact on paper was like trying to pick up a bead of quicksilver with our fingertips.

We believe, however, that by sticking to indisputable scientific facts and treating new developments in a historical perspective, we have been able to provide much of the information that people need to make intelligent, informed decisions about radiation exposure.

In the last two decades of the twentieth century, we will make some of the biggest decisions ever demanded of a nation. And, as always, these decisions will require us to weigh benefits versus risks.

In the area of nuclear power, are the risks of an accident more significant than the benefits afforded by a plentiful, inexpensive, and dependable supply of clean energy that we needn't import? Is nuclear power a monstrous threat that jeopardizes the entire human race? Or is it a boon that will save us from a dark age brought about by dwindling energy resources?

Are people who feel that nuclear energy represents our salvation really irresponsible powermongers who are ready to sacrifice the earth on an atomic altar? Are the people who oppose nuclear energy really just a bunch of directionless hippies looking for the polarizing cause they have lacked since Vietnam? Isn't there a sensible position somewhere between these opposite ends of the spectrum? Can't we enjoy the benefits of nuclear power safely? The answer lies in the willingness of the people and their leaders to deal not in emotional posturing but in solid facts.

The most important of these facts is that nuclear power is much too precious a resource to toss aside lightly. But make no mistake, our survival does not depend on it. At present, only about 13 percent of our electricity is generated in nuclear power plants, and we can produce much more power than we actually need. If such plants were proved to be an intolerable threat, we could manage without them —but not easily. Nuclear energy is cleaner, less expensive, and not dependent on a commodity that can be withheld from us. Perhaps the wisest course at this moment in history is to invest our energy hopes (and our energy dollars) in two sources, coal and nuclear fuel. Then, if the risks involved in producing nuclear energy become too heavy to bear, we can phase it out and still survive.

Let us briefly review those risks. First, of course, is radiation hazard. But remember that, under ordinary circumstances, nuclear power plants are *not* radiation hazards. In fact, they contribute less radioactivity to the environment than coal-fueled power plants.

Another source of concern is the transportation of nuclear materials. Yet such materials are shipped in containers that can weather the shocks they might be subject to in a fall or a traffic accident. If small risks do exist, they can be further reduced by changing routes and times of movement.

Then there is the risk of strategic arms proliferation and the possibility that nuclear weapons will fall into irresponsible hands. While this is a legitimate cause for concern, we have come too far down the road to change anything, even if we unilaterally dismantled all our nuclear weapons, which would be foolhardy indeed.

Disposal of radioactive waste material entails yet another risk. At present, such waste is being stored safely, if only temporarily. There is no immediate rush to move it. Admittedly, solutions will eventually be needed. But most authorities agree that, given the state of today's technology, solutions will be found.

And, of course, there is the inevitable risk of blunders by nuclear power plant technicians, such as occurred at Three Mile Island. Current efforts to upgrade the training of nuclear power plant personnel in emergency procedures should be applauded. But, in view of the *potential* for disaster, it would be prudent to locate these plants so far from major metropolitan areas that no accident could threaten large numbers of people trapped because evacuation was not feasible.

And what of radiation in the medical arts? Here, too, we have talked at great length about benefits and risks. In this area, the risks are simpler to understand. Overexposure to radiation, regardless of how it occurs, carries with it the risk of cancer, birth defects, and genetic mutations. But, like nuclear energy, radiation is too valuable a medical tool to discard, *unless* it turns out to be more destroyer than savior.

Again, the most prudent course lies somewhere between the extremes of outlawing all radiation procedures and indiscriminately irradiating every patient who comes anywhere near a doctor's or dentist's office. We have offered some ideas about what can be done to maximize benefit and minimize risk. Clearly, more work is needed in this critical area. But with the accelerating pace of medical research, there is every reason to hope that new technological breakthroughs will provide the answers we all seek.

Finally, what should be done about all those devices that so improve the quality of our lives, yet expose us to low-level radiation and possible biological damage? Here again, sensible options exist that permit microwave, infrared, and ultraviolet appliances to be used and enjoyed with little risk. Furthermore, intensive research programs, begun in response to the challenges of the space age, promise even greater improvement of the balance.

In compiling this radiation primer, we have tried to present an overview of the subject from all the angles you need to find your own perspective. Now it's time for you to decide whether to stop worrying . . . or to start.

Glossary

Acute radiation sickness The immediate effects, such as nausea and loss of appetite, that follow exposure to high levels of penetrating radiation

Alpha ray A stream of charged particles made up of two protons and two neutrons from the nucleus of an atom

Angiography The visualization of blood vessels through use of x-rays applied during the injection of an iodine-containing liquid

Angioma A tumor composed of blood or lymph vessels

Antiparticle A constituent of antimatter—the form of matter in which the electrical charge of particles is reversed from conventional matter

Atoms Any of the smallest particles of an element that combine with other elemental particles to form compounds

Atomic number A number reflecting the positive charge or quantity of protons in the nucleus of an atom of a particular element

Atomic pile A device that houses a nuclear chain reaction used to produce atomic energy

Beta ray A stream of particles from the nucleus of an atom, with a mass and charge equal to an electron

Betatron A device for accelerating electrons to an energy level higher than one billion electron-volts

Biologic half-life The time it takes for a living organism to eliminate half the amount of radioactive substance it has absorbed

Blepharitis Inflammation of the eyelids

Brachytherapy A form of radiation therapy in which the source is close to the area of the body being treated

Breeder reactor A nuclear reactor that not only generates energy but creates more fissionable material than it utilizes

Cadmium A bluish-white metallic element with an atomic number of 48

Calcine To incinerate to ashes

Carcinogenic Capable of producing cancer

Cardiac pacemaker A device implanted in the body that stimulates contraction of the heart muscle at a fixed rate by electrical stimulation

Chain reaction A series of chemical reactions in which materials are activated and cause additional reactions to occur

Chromosome A structure carrying the genes that transmit genetic information and govern heredity

Computerized axial tomography (CAT) A procedure in which x-rays are applied to thin layers of tissue, and are organized by a computer into pictures of anatomic slices of different body parts.

Control rod A bar usually made of boron steel that can be inserted into an atomic pile to absorb neutrons, and slow down or stop a nuclear chain reaction

Cosmic ray A stream of charged particles with short wavelengths and great penetrating power originating in outer space

Cosmotron A device that accelerates protons to extremely high energy levels

Critical mass The amount of radioactive substance large enough so that exactly one neutron from each reaction causes a further reaction

Cyclotron An apparatus for accelerating protons to high energies

Deoxyribonucleic acid (DNA) An organic substance found in chromosomes that carries genetic information

Deuterium An isotope of hydrogen having an atomic weight of approximately 2

Down's syndrome A condition caused by a specific alteration in a chromosome; often mistakenly called mongoloid idiocy

Electrolyte A substance which when dissolved in a fluid permits the passage of electricity through that fluid

Electromagnetic field The space in which exists the effects of magnetism caused by a current of electricity

Electromagnetism An attraction that results from the passage of an electrical current

Electron An elementary particle charged with negative electricity

Electron volt A unit of energy equivalent to the energy gained by an electron passing through a difference in potential of one volt

Energy The capacity for performing a physical action

Enrichment The process of artificially increasing the percentage of an isotope beyond that which is normally present in a given material

Excitation The process by which the energy of a molecule, atom, electron, or nucleus is increased beyond the normal amount

Fallout Radioactive, airborne particles that reach the ground after a nuclear explosion

Fertilized ovum A female egg cell that has undergone impregnation by the male reproductive cell

Fission The splitting of a nucleus with resultant release of large amounts of energy

Fluorescence Emission of electromagnetic radiation, usually as visible light

Fluoroscope An instrument for visualizing internal structures of the body through the use of x-rays

Fluoroscopy Examination by means of a fluoroscope

Food-chain pathways The route by which radioactive material or any other substance finds its way into the digestive tract of man

Frequency The number of repetitions of a process, such as electromagnetic waves, within a given unit of time

Fuel rod A bar of fissionable material that can be used in a nuclear reactor to produce energy

Fusion The formation of a heavy nucleus by the joining together of two lighter ones, with the production of energy

Gamma ray Electromagnetic radiation of high energy and short wavelength

Geiger counter An instrument used to detect and measure radioactivity by means of its passage through a gas-filled tube containing electrodes that react to ionizing radiation

Gene A protein molecule found in the chromosomes that transmits heredity

Genetic effects Effects that can be transferred from parent to offspring through alteration of the genes

Gestation The period of pregnancy

Half-life The period of time in which half the atoms of a radioactive substance decay to a different nuclear form

Heat exchanger The part of a nuclear reactor that transfers heat produced during fission to ordinary water to form the steam that drives turbines

Heavy hydrogen An isotope of hydrogen with a nucleus containing one proton and one neutron; also called deuterium

High-level waste Discarded material that is still considerably radioactive

High temperature reactor A reactor that produces temperatures high enough to generate mechanical power efficiently using gas as a cooling medium

Histology The study of the minute structures of plant or animal tissue

Hydrocephaly An abnormal condition caused by increased amounts of fluid within the brain

Hyperbilirubinemia Excessive amounts of bile pigment found in the blood

Infrared radiation Nonionizing thermal radiation from the red end of the spectrum, of wavelengths longer than those of visible light

Ion An atom or group of atoms carrying a positive or negative charge of electricity, due to the loss or gain of one or more electrons

Ionization chamber A partially evacuated tube used to detect and measure ionizing radiation by quantifying the flow of electricity through the remaining gas, which undergoes ionization

Ionizing radiation A type of radiation in which electrons are expelled from atoms or molecules, thus producing ions

Ionosphere The section of the earth's atmosphere from an altitude of 25 miles to about 250 miles or so, that contains free ions and thus permits the transmission of radio waves

Isotope An element having the same atomic number, but differing in the number of neutrons in its nucleus

Kilowatt A unit of electric power equal to 1,000 watts, where one watt corresponds to 1/746 horsepower

Laser An acronym formed from the first letters of *l*ight *a*mplification by *s*timulated *e*mission of *r*adiation; actually a device that uses the properties of specific molecules to produce the emission of light radiation as a powerful directional beam

Latent period The span of time between exposure to a disease-producing agent and response

Leukemia A blood disease characterized by the presence of an abnormal number of white blood cells

Light water reactor A widely used type of reactor in which ordinary light water slows down the neutrons produced and then transfers the heat of the reaction

Linear accelerator A long, tubelike structure in which the flow of electrons or protons is speeded up by means of oscillating electromagnetic fields, thereby increasing the energy of the particles

Linear energy transfer (LET) A measure of the capacity of biological tissue to absorb radiation; actually, the amount of energy lost as radiation passes along a path of specific length through tissue

Liquid metal fast breeder reactor A reactor in which plutonium and uranium are immersed in liquid sodium, which absorbs the heat of the reaction and carries it to a heat exchanger system

Low level waste Discarded radioactive material that is contaminated by only moderate quantities of mixed fission products

Lymphocyte A white blood cell formed in the lymphatic tissue of the body

Magnetopause The outer edge of the magnetosphere, a region of the earth's atmosphere extending thousands of miles into space; the magnetosphere comprises the earth's magnetic field, in which charged particles are trapped

Mammogram A picture of the tissues of the breast produced by x-rays

Mass The amount of matter contained in a body; the term is often used synonymously with weight

Melanoma A tumor, usually malignant, containing dark pigment

Meltdown A nuclear reactor accident in which control of the nuclear reaction is lost and heat is generated in such quantities as to cause the entire case and its housing to dissolve

Meson Any of a group of unstable nuclear particles with a mass between that of an electron and a proton

Metastasis The process in which a disease is transferred from a localized diseased tissue to a different area of the body

Microcephaly An abnormal condition characterized by small head size and usually associated with mental defects

Microwave An electromagnetic wave whose length falls somewhere between a few tenths of a millimeter and about thirty centimeters

Mitosis A process that occurs within the nucleus of a dividing cell, resulting in the development of two new nuclei

Moderator In a nuclear reactor, any substance used to slow down neutrons

Molecule The smallest particle of a substance, containing one or more atoms, that retains the characteristics of the substance

Muon A contraction of mu-meson; a negatively or positively charged unstable elementary particle common in cosmic radiation, with a mass about 200 times greater than the electron

Mutagen A substance that causes mutations to occur

Mutation A significant alteration in the characteristics of an organism caused by alterations in the structure of chromosomes or genes

Neoplasm A growth of new tissue that serves no physiologic purpose

Neutrino An uncharged elementary particle of little mass but high penetrating power that weakly interacts with matter and therefore easily escapes detection

Neutron An elementary particle that is electrically neutral, has a mass approximately that of the proton and exists in every known atomic nucleus except that of hydrogen

Nonionizing radiation A type of electromagnetic radiation that does not cause the displacement of electrons from atoms or molecules

Nuclear fuel cycle The series of stages in providing nuclear reactor fuel, from mining of the ore through refabrication into new fuel elements and management of radioactive waste

Nuclear waste Materials from nuclear operations that are contaminated with radioactivity and have no further use

Nuclear reactor A piece of equipment in which a fission chain reaction can be started, maintained, and controlled to generate energy

Nuclide A species of atom with a specific charge, mass, and energy state, and capable of existing for a measurable lifetime.

Nucleus The central part of an atom, the basic components of which are protons and neutrons; the nucleus makes up most of the mass of the atom

Oligospermia A deficiency of sperm cells in the male semen

Oncologist A physician who specializes in the study of tumors

Phosphor A material that gives off light when exposed to radiation

Phosphorescent Able to admit light without noticeable heat or combustion

Photon A quantum of light energy that is electrically neutral and has no mass

Pion A contraction of pi-meson; a negatively or positively charged, or neutral, elementary particle with a mass slightly larger than a muon and about 270 times greater than an electron

Pitchblende An ore that contains uranium compounds

Plasma In physics, a hot, neutral gas containing positive and negative ions; in biology, the fluid part of blood

Positron The positive counterpart of an electron, also known as an antielectron

Pressurized water reactor A type of reactor in which water does not boil, but is pumped under pressure at high temperatures through a heat-exchanging device that transfers its heat to ordinary water

Prodrome A warning symptom or symptoms that signal the onset of an illness

Proton A particle with a single positive electrical charge found in the nucleus of an atom

Quality assurance (QA) Actions taken to assure that x-rays will be of high quality with minimum exposure to patient or operator

Quantum An elemental unit of energy

Quark A theoretical grouping of elementary particles

Rad From the first letters of *r*adiation *a*bsorbed *d*ose; the unit that denotes absorbed dosages of ionizing radiation

Radiation The process in which atoms and molecules undergo internal change, with the resultant emission of energy as streams of fast-moving particles or light

Radioactive Capable of emitting ionizing radiation through the spontaneous disintegration of atomic nuclei

Radioactive series A succession of isotopes of various elements formed by radioactive disintegration until a stable state is reached

Radiology The science dealing with the use of ionizing radiation for diagnosing or treating various diseases

Radiometer An instrument used to detect and measure the intensity of radiant energy

Radiopharmaceutical A drug or chemical rendered radioactive to fulfill a specific function

Radon daughter A nuclide formed by the radioactive decay of the element radon, itself a daughter of radium

Referral criteria A series of guidelines to be followed when consideration is being given to whether a patient needs diagnostic x-rays

Relative biological effectiveness (RBE) A factor used to compare the effects of absorbed radiation doses due to different types of ionizing radiation

Rem From the first letters of roentgen-equivalent-man; the unit of ionizing radiation dose that has the same biological effects as one rad of x-rays

Roentgen A unit used to measure x-ray radiation exposure

Scintillation counter An instrument for detecting and measuring radioactivity by means of a photoelectric cell that converts radiation into light flashes

Solar flare A sudden and temporary explosion of energy emanating from a limited area of the sun's surface

Somatic effect The consequences of radiation directly upon the individual exposed rather than on his offspring

Strip mining Excavation of ore after exposing it through removal of the covering layers of the earth's surface

Sun spots Dark areas on the sun's surface believed to be tornadolike solar storms

Synchrocyclotron A type of device for accelerating the velocity of charged particles to produce tremendous amounts of energy

Synchotron A more sophisticated type of synchrocyclotron and capable of producing greater amounts of energy

Technologically enhanced natural radiation (TENR) A term denoting actions taken by man that increase the radioactivity of a natural substance or specific locale

Thermography A technique in which an infrared sensitive camera is used to show surface temperatures of the body photographically as a means of diagnosing underlying tumors

Thermonuclear reaction One which depends on extremely high temperatures to fuse two light nuclei, forming the nucleus of a heavier atom with the release of enormous energy

Thimble chamber An ionization chamber for measuring roentgens

Tomography A special x-ray technique for showing detailed images of structures lying in just one plane of tissue

Transuranic Designating elements with atomic numbers higher than uranium, such as plutonium and neptunium

Tritium A radioactive isotope of hydrogen containing two neutrons and one proton in the nucleus

Troposphere The atmosphere below the stratosphere in which clouds form and temperature decreases with altitude

Turbogenerator A device for producing energy, coupled directly to an engine or motor driven by the pressure of steam, water, or air

Ultraviolet radiation Light rays of extremely short wavelength which emanate from just beyond the violet end of the visible spectrum

Uranium A hard, heavy, radioactive metallic chemical element found naturally as a compound in ores such as pitchblende

Van Allen belt Either of two zones of intense natural radiation that surround the earth at various altitudes in the upper atmosphere

Video display terminal (VDT) A screen on which can be visualized information retrieved from a computer

Wavelength The distance between corresponding points on two successive waves

Xerophthalmia A medical condition characterized by dryness of the eye

X-ray An electromagnetic radiation of extremely short wavelength capable of penetrating opaque or solid tissue

Selected Bibliography

A. NATURALLY OCCURRING IONIZING RADIATION

Cesium-137 from the Environment to Man: Metabolism and Dose. National Council on Radiation Protection and Measurement Report 52, January 1972.

Energy, the Environment and Human Health. American Medical Association, Publishing Sciences Group, 1973.

Environmental Radiation Measurements. National Council on Radiation Protection and Measurement Report 50, December 1976.

J. Lavernhe, E. LaFontaine, and R. LaPlane, "The Concorde and Cosmic Rays." *Aviation, Space and Environmental Medicine,* February 1978, pp. 419–421.

Natural Background Radiation in the United States. National Council on Radiation Protection and Measurement Report 45, November 1975.

Radiologic Quality of the Environment in the United States, 1977. National Technical Information Service, PB-274229, September 1977.

B. NUCLEAR ENERGY

Energy, the Environment and Human Health. American Medical Association, Publishing Sciences Group, 1973.

Nuclear Power at Northeast Utilities: Fact Book 1979. System Communications Department of Northeast Utilities, Hartford, Connecticut.

Risks Associated with Nuclear Power: A Critical Review of the Literature. National Academy of Sciences, 1979.

C. BIOLOGICAL EFFECTS OF IONIZING RADIATION

M. H. Barnett, *The Biological Effects of Ionizing Radiation: An Overview.* HEW (FDA) Publication 77-8004, October 1976.

Beehrs, Shapiro, Smart et al., "Working Group to Review the Breast Cancer Detection Demonstration Projects." *Journal of the National Cancer Institute,* Vol. 62, No. 3 (March 1979), pp. 644–695.

J. D. Boice, C. E. Land, R. E. Shore, J. E. Norman, and M. Tokunaga, "Risk of Breast Cancer Following Low-Dose Radiation Exposure." *Radiology,* Vol. 131 (June 1979), pp. 589–597.

I. D. J. Bross et al., "Accumulative Genetic Damage in Children Exposed to Pre-Conception and Intrauterine Radiation." *Investigative Radiology*, Vol. 15, No. 1 (January–February 1980), pp. 52–64.

R. P. Chiacchierini, F. E. Lundin, and P. C. Scheidt, "A Risk-Benefit Analysis by Life Table Modeling of an Annual Breast Cancer Screening Program Which Includes X-Ray Mammography." *Proceedings of the Third International Symposium on the Detection and Prevention of Cancer*, New York, New York, April 26–May 1, 1976.

R. A. Conard et al., "Thyroid Neoplasia as Late Effect of Exposure to Radioactive Iodine in Fallout." *Journal of the American Medical Association*, Vol. 214, No. 2 (October 1970) pp. 316–324.

M. J. Dwyer and D. B. Leeper, *Carcinogenic Properties of Ionizing and Non-Ionizing Radiation*, Vol. 3: *Ionizing Radiation*. DHEW (NIOSH) Publication No. 78-142, April 1978.

"Effects of Atomic Radiation." *United Nations Review*, Vol. 9, No. 10.

Effects on Populations of Exposure to Low Levels of Ionizing Radiation. Report of the Advisory Committee on the Biological Effects of Ionizing Radiation, National Academy of Sciences, November 1972.

Effects on Populations of Exposure to Low Levels of Ionizing Radiations. Report of the Committee on the Biological Effects of Ionizing Radiation, Criteria and Standards Division, Office of Radiation Programs, U.S. Environmental Protection Agency, 1979.

Gonad Doses in Genetically Significant Dose from Diagnostic Radiology, U.S., 1964–1970. Department of Health, Education and Welfare, April 1976.

A. P. Hufton, "Radiation Dose to the Fetus in Obstetric Radiography." *British Journal of Radiology*, Vol. 52 (September 1979), pp. 735–740.

J. L. Lyon, M. R. Klauber, J. W. Gardner and K. S. Udall, "Childhood Leukemias Associated with Fall-Out from Nuclear Testing." *New England Journal of Medicine*, Vol. 300 (February 22, 1979), pp. 397–402.

J. H. MacGregor et al. "Breast Cancer Incidence Among Atomic Bomb Survivors, Hiroshima and Nagasaki 1950–1969." *Journal of the National Cancer Institute*, Vol. 59 (1977), pp. 799–811.

R. H. Mole, "Radiation Effects on Prenatal Development and Their Radiological Significance." *British Journal of Radiology*, Vol. 52, No. 614 (February 1979), pp. 89–101.

C. E. Moss et al., *Manual of Health Aspects of Exposure to Non-Ionizing Radiation*. 1979.

National Academy of Sciences, *Biological Effects of Atomic Radiation*, Summary Reports 1956.

S. M. Neuder, *Electromagnetic Fields in Biological Media*. DHEW (FDA) Publication 79-8072, August 1979.

B. Shleien et al., "Mean Active Bone Marrow Dose to the Adult Population of the United States from Diagnostic Radiology." *Health Physics,* Vol. 34 (June 1978), pp. 587–601.

C. Silverman and D. H. Hoffman, "Overview—Thyroid Tumor Risk from Radiation During Childhood." *Preventive Medicine,* Vol. 4 (1975), pp. 100–105.

E. L. Socolow et al., "Thyroid Carcinoma in Man After Exposure to Ionizing Radiation: A Summary of the Findings in Hiroshima and Nagasaki." *New England Journal of Medicine,* Vol. 268, No. 8 (February 1963), pp. 406–410.

Sources and Effects of Ionizing Radiation. United Nations Scientific Committee on the Effects of Atomic Radiation, 1977.

D. ULTRAVIOLET RADIATION

T. P. Coohil, *Wavelength Dependence of Ultraviolet Enhanced Reactivation and Induction of Mammalian Viruses.* DHEW (FDA) Publication 78-8059, March 1978.

Mercury Vapor in Metal Halide Lamps. FDA Consumer Memo, DHEW (FDA) Publication 77-8041, November 1977.

L. F. Mills et al., *A Review of Biological Effects and Potential Risks Associated with Ultraviolet Radiation as Used in Dentistry.* DHEW (FDA) Publication 76-8021, October 1975.

E. LASERS

F. Alan Andersen, *Biological Bases for and Other Aspects of a Performance Standard for Laser Products.* DHEW (FDA) Publication 75-8004, July 1974.

Laser and Optical Hazards Course Manual. U.S. Army Environmental Hygiene Agency, January 1979.

R. W. Peterson, *Some Considerations of Hazards in the Use of Lasers for Artistic Displays.* DHEW (FDA) Publication 79-8082, March 1979.

W. F. VanPelt, H. F. Stewart, R. W. Peterson, A. M. Roberts, and J. K. Worst, *Safety in Classroom Laser Use.* DHEW Publication, May 1970.

F. LIGHT ENERGY

R. E. Behrman, "Preliminary Report of the Committe on Phototherapy in the Newborn Infant." *Fetal and Neonatal Medicine,* Vol. 84, No. 1 (1974), pp. 135–147.

S. Cunningham-Dunlop and B. H. Kleinstein, *A Current Literature Report on the Carcinogenic Properties of Ionizing and Non-Ionizing Radiation. I. Optic Radiation.* DHEW (NIOSH) Publication 78-122, December 1977.

K. R. Endall and E. J. Shangold, *Survey of Photocopier and Related Products.* DHEW (FDA) Publication 78-8060, March 1978.

D. G. Hazzard, *Symposium on Biological Effects and Measurements of Light Sources.* DHEW (FDA) Publication 77-8002, October 1976.

P. C. Scheidt et al., "Toxicity to Bilirubin in Neonates: Infant Development during First Year in Relation to Maximum Neonatal Serum Bilirubin Concentration." *Journal of Pediatrics,* August 1977, pp. 292–297.

G. MICROWAVE RADIATION

Biologic Effects and Health Hazards of Microwave Radiation. Proceedings of an International Symposium, Warsaw, October 1973. Polish Medical Publishers.

Biological Effects of Microwave Radiation. Institute of Electrical and Electronics Engineers, January 1980.

M. J. Dwyer and D. B. Leeper, *A Current Literature Report on the Carcinogenic Properties of Ionizing and Non-Ionizing Radiation, Microwave and Radiofrequency Radiation.* DHEW (NIOSH) Publication 78-134, March 1978.

A. J. N. Nelson et al., "Combined Microwave Therapy." *Medical Journal of Australia,* Vol. 2 (1978), pp. 88–90.

Radiofrequency Microwave and Ultrasound Hazards. Course Manual, U.S. Army Environmental Hygiene Agency, 1971.

"Symposium on Health Aspects of Non-Ionizing Radiation." *Bulletin of the New York Academy of Medicine,* Vol. 55, No. 11 (December 1979).

H. REDUCING RADIATION

"Diagnostic X-Ray Systems Amendment of Assembly and Reassembly Provision and Performance Standard (DHEW)." *Federal Register,* Vol. 44, No. 166 (August 1979), pp. 49667–49672.

L. W. Goldman, *Analysis of Retakes: Understanding, Managing, and Using an Analysis of Retakes Program for Quality Assurance.* DHEW (FDA) Publication 79-8097, August 1979.

"Quality Assurance in Nuclear Medicine; Intent to Propose Voluntary Recommendations (DHEW)." *Federal Register,* Vol. 44, No. 161 (August 1979), pp. 48264–48265.

"Quality Assurance Programs for Diagnostic Radiology Facilities (DHEW)." *Federal Register,* Vol. 44, No. 239 (December 1979), pp. 71728–71740.

Radiation Exposure from Consumer Products and Miscellaneous Sources. National Council on Radiation Protection and Measurement Report 56, November 1977.

J. R. Skalnic, "Carbon Fiber and Dose Reduction." *Radiation Today,* Vol. 2, No. 4 (September/October 1980), pp. 40–42.

J. R. Skalnic, "Radiation Reduction." *Radiology Today,* December/January 1980, pp. 6–19.

W. M. Whitehouse, C. S. Simons and T. N. Evans, "Reduction of Radiation Hazard in Obstetric Roentgenography." *American Journal of Roentgenology, Radium Therapy, and Nuclear Medicine*, Vol. 80, No. 4 (October 1958), pp. 690–695.

I. MEDICAL AND DENTAL USES OF RADIATION

R. A. Brooks and G. DiChiro, "Principles of Computer Assisted Tomography (CAT) in Radiographic and Radioisotopic Imaging." *Physics, Medicine, Biology*, Vol. 21, No. 5 (1976), pp. 689–732.

R. F. Brown, J. W. Shaver and D. A. Lamel, "A Selection of Patients for X-Ray Examination." DHEW (FDA) Publication 80-8104, January 1980.

R. P. Chiacchierini and F. E. Lundin, Jr., "Risk Benefit Analysis for Reduced Dose Mammography." From *Reduced Dose Mammography*, Wendy W. Logan and E. Philip Muntz, eds. New York: Masson Publishing, 1979.

"Dental Post Treatment Radiographs; Proposed Recommendations." *Federal Register*, Vol. 44, No. 138 (July 1979), pp. 41486–41487.

The Developing Role of the Short-lived Radionuclides in Nuclear Medicine. DHEW (FDA) Publication 77-8035, August 1977.

"Diagnostic X-Ray Systems and Their Major Components; Amendments to Performance Standards (DHEW)." *Federal Register*, Vol. 44, No. 151, (August 1979), pp. 45645–45647.

National Conference on Referral Criteria for X-Ray Examinations. DHEW (FDA) Publication 79-8083, April 1979.

National Conference on Referral Criteria for X-Ray Examinations, October 25–27, 1978. DHEW (FDA) Publication 79-8083, April 1979.

L. A. Phillips, *A Study of the Effect of High Yield Criteria for Emergency Room Skull Radiography.* DHEW (FDA) Publication 78-8069, July 1978.

"The Risks of Mammograms (Commentary)." *Journal of the American Medical Association*, Vol. 237, No. 10 (March 7, 1977), pp. 965–966.

M. Rosenstein, *Handbook of Selected Organ Doses for Projections Common in Diagnostic Radiology.* DHEW (FDA) Publication 76-8031, May 1976.

E. L. Saenger, "Radiologists, Medical Radiation and the Public Health." *Radiology*, Vol. 92 (March 1969), pp. 685–699.

B. F. Wall, D. A. C. Green, and R. Veerappan, "The Radiation Dose to Patients from EMI Brain and Body Scanners, 1979." *British Journal of Radiology*, Vol. 52 (1979), pp. 189–196.

V. F. Wall, E. S. Fisher, R. Paynter, A. Hudson, and P. D. Bird, "Dosage to Patients from Pantomographic and Conventional Dental Radiography." *British Journal of Radiology*, Vol. 52 (1979), pp. 727–734.

J. NUCLEAR ACCIDENTS

Population Dose and Health Impact of the Accident at Three Mile Island Nuclear Station. Ad Hoc Population Dose Assessment Group, May 1979.

Population Dose and Health Impact of the Accident at the Three Mile Island Nuclear Station (A Preliminary Assessment for the Period March 28 through April 7, 1979), May 10, 1979. HFX-25, Bureau of Radiological Health.

K. RADIATION PROTECTION

Atomic Energy Research. United States Energy Commission, 1961.

Basic Radiation Protection Criteria. National Council on Radiation Protection and Measurement Report 39, 1971.

Gonad Shielding in Diagnostic Radiology. DHEW (FDA) Publication 75-8024, June 1975.

Instrumentation and Monitoring Methods for Radiation Protection. National Council on Radiation Protection and Measurement Report 57, May 1978.

M. B. Mahaffey and R. B. Lewis, "Radiation Safety." *Veterinary Medicine,* August 1979, pp. 9–13.

Manual on Radiation Protection in Hospitals. World Health Organization, 1974.

Medical Radiation Exposure of Pregnant and Potentially Pregnant Women. National Council on Radiation Protection and Measurement Report 54, July 1977.

Protection of the Thyroid Gland in the Event of Releases of Radioiodine. National Council on Radiation Protection and Measurement Report 55, August 1977.

"Quality Assurance in Nuclear Medicine; Intent to Propose Voluntary Recommendations (DHEW)." *Federal Register,* Vol. 44, No. 161 (August 1979), pp. 48264–48265.

"Quality Assurance Programs for Diagnostic Radiology Facilities (DHEW)." *Federal Register,* Vol. 44, No. 239 (December 1979), pp. 71728–71740.

Radiation Protection for Medical and Allied Health Personnel. National Council on Radiation Protection and Measurement, 1976.

Review of the NCRP Radiation Dose Limits for Embryo and Fetus in Occupationally Exposed Women. National Council on Radiation Protection and Measurement Report 53, November 1976.

L. NUCLEAR WASTE DISPOSAL

R. P. Hammond, "Nuclear Waste and Public Acceptance." *American Scientist,* Vol. 67 (March–April 1979), pp. 146–150.

M. MEASURING RADIATION

M. Rosenstein, *Organ Doses in Diagnostic Radiology.* Department of Health, Education and Welfare, May 1976.

Symposium on Biological, Imaging Techniques, Dosimetry of Ionizing Radiations. HHS (FDA) Publication 80-8126, July 1980.

Index

absorption: of microwave energy, 57; safe levels of, 124; of x-rays, 104–105
accelerators, particle, 27–28
accidents, nuclear, threat of, 85, 89–90, 91
activation analysis, radiation and, 149–50
acute radiation sickness, 106–108; dosage of radioactivity and, 107–108; Hiroshima and Nagasaki bombs, knowledge from, 106–107; somatic effects of, 106; symptoms of, 107
age, as variable in radiation doses, 52–53; rad/rem correlations and, 109–10
air travel, radiation and, 37, 70
ALARA safety standard of radiation, 195
alpha particles, 12, 26, 43
American Cancer Society, 139, 142
analysis, activation, radiation and, 149–50
animal(s): radioactivity from eating, 43; studies, microwave and radar radiation and, 124–26
antielectron (positron), 8
antiproton, 8
artificial production of radioactivity, by particle accelerators, 27–28
atmosphere, as protection against cosmic rays, 36
atom(s), 3–8; antiparticle, 8; bomb, 4, 78; electricity and, 4–5; electrons, 4, 5; electromagnetism and,

7; energy and mass, 4; gravity and, 7; helium, 6; hydrogen, 6; hyperons, 8; interaction between, strong and weak, 7; isotopes and, 6; lithium, 6; mesons, 7–8; neutrino, 8; neutron, 6; nucleus of, 4, 5–6; omega-minus particle, 8; particles and quarks, families of, 8; photons, 7; protons, 7; subatomic particles, 5, 7; types of, 6
atoms, restlessness of, 9–19; chemical activity, 9–11; electrical, 11; electromagnetic radiation (energy), 11, 13, 14–19; radioactivity, 12–14
Atomic Energy Commission, 66, 87, 89
average human dose of radiation, lifetime, 47–48

baryons, 8
beta particles, 12, 26–27
Bev (billion electron volts), 29
biological damage from ionizing radiation, 104–105; linear energy transfer (LET), 32, 104
birth defects, ionizing radiation and, 109, 111
bismuth, 20, 21
Bloomfield, Colorado, leak of nuclear wastes at, 99
body, human: area exposed to ionizing radiation, 105–106; bone marrow exposure to radiation, estimated, from x-rays (chart), 51;

CAT scan, for medical diagnosis, 140–42; infrared radiation, 131–33; ionizing radiation on, estimating (chart), 68; lasers and, 133–34, 135–36; liquids, radioactive, problem of, 95–96; microwave radiation in, safe levels of, 124, 126–30; organs, target and critical, in radiopharmaceuticals, 52; potassium in, 25; radiopharmaceuticals for diseases, 144; ultraviolet light, 130–31; "whole radiation" in, external and internal, 39, 42, 46; x-rays, for medical diagnosis and treatment, 137–40, 142–43. See also health-related radiation exposure; tissue, human

Bohlen, Charles, 129

bombs: atomic, 4; fission, 79, 80; hydrogen, 79, 147

bone(s): cancer of, 120; marrow, radiation effects on, 49–51; and teeth, composition of, radiation and, 17

brachytherapy, 144

breast cancer, 120–21, 139. See also cancer

breeder reactors, 85

Bureau of Radiological Health, 60, 175, 177, 178, 185–87, 193

burial of nuclear wastes, locations of, 98, 100–101

cadmium-plated rods, 80

calcium, internal radiation and, 43

Califano, Joseph A., Jr., 103

cancer, 49–50, 142; brachytherapy for, 144; laser therapy for, 144–45; mammography and, 120–22, 139–40, 158; nonionizing radiation and, from microwaves, 128–30; radiation treatment for, 48–49; skin, 120, 130–31, 132, 164–65; of thyroid, 53; x-ray therapy for, 142–43

cancer, as effect of ionizing radiation, 114–22; bone cancer, 120;

breast cancer, 120–22, 139; carcinogen, radiation as, 117–19; cause-effect relationships, 114; digestive system cancer (stomach), 120; leukemia, 118, 119; lung cancer, 120; populations, particular, cancers linked to radiation in (chart), 115–16; research difficulties, 114, 117; skin cancer, 120; tissue damage, radiation dosage and, 122; thyroid cancer, 119. See also mutation

Carter, Jimmy, xiii

carbon-14, 45, 66, 86

carcinogen, radiation as, !17–19; evidence for, 118; Hiroshima and Nagasaki survivors and, 118; risk estimate in populations, 119; theories about, 117–18

cardiac pacemakers. See pacemakers

CAT scanners (computerized axial tomography), 49, 159; for medical diagnosis, 140–42. See also x-rays

cells, body, radiation and: biology of, genetic effects of radiation and, 112–13; ionizing radiation, coping with, 106–107; mutations from radiation exposure, 113–14; theory of chromosomal damage, 117. See also body, human; tissue, human

Center for Disease Control, 187

cereals, radioactivity in, 43

cesium: -136, nuclear testing and, 66; -137, as nuclear waste, 86, 96

chain reaction, fission, 78

chambers, radiation, 30–31

charges, positive and negative electrical, 4–5; ions and, 10–11

chemical activity, atoms and, 9–11; compounds, 10; molecular motion, 10; molecules, 10

chemicals, hereditary apparatus and, 112–13

children: radiation exposure and, 103, 109–10; x-rays and, 157

China syndrome. See "meltdown"

chloride ions, 10

chromosomes. *See* cells, body

cigarette smoking, radiation from, 46, 161

civilian applications of radiation, 148–50; activation analysis, 149–50; communications, 148; food preservation, 148–49; industrial uses, 150; insect control, 150; radar, 149

classifications of nuclear wastes and corresponding problems, 94–97; distinctions in, 94; facilities and equipment, 95; liquids, radioactive, 96–97; spent fuel problem, 95–96; transuranic elements, 95. *See also* wastes

coal, 146

cobalt-60, 101, 143

collimation of dental x-rays, 160

Commerce, Department of, 189

Commonwealth Edison Dresden II nuclear accident, 89

communications, use of radiation in, 148

compounds, chemical, 10

computer contributions to radiation safety, 175–76

computerized axial tomography. *See* CAT scanners

congenital defects, radiation and, 109

Congressional Office of Technological Assessment, 146–47

construction materials, radioactivity and, 41, 46

Consumer Product Safety Commission (CPSC), 165, 191–92

consumer products, x-ray safety and, 69–70, 160–63; cigarette smoking, 161; lasers, 135–36, 137; microwave ovens, 168–71; nuclear reactors, 161–63; televisions, 161; tungsten halide bulbs, 166–67; watches and clocks, 59

control rods, in nuclear reactor, 81

cooling fluid, in nuclear reactor, 81

cosmic rays, 35–39; atmosphere, as protection against, 36; external exposure of body to (chart), 39; geography and exposure to, 37–38; intensity of, altitudes and, 36–37; ionosphere, as protection against, 38; magnetopause, as protection against, 38; primary and secondary radiation and, 35; from stars, 36; from sun, 36; from Van Allen belts, 36. *See also* gamma rays

Cosmotron, 29

counters, for radiation chambers, 30–31

critical mass, in fission bombs, 78, 80

critical organs, radiopharmaceuticals and, 52

crystalline material in scintillators, 30–31

cyclotron, 28

deaths in nuclear industry, 88, 90, 91

decay, radioactive, 20–25; energy production through, 22; half-life, 23; isotopes, radioactive, of stable elements, 25; radioactive series, 21–22; uranium (radium) series (chart of), 24; uranium-238, odyssey of, 21

Defense, Department of, 188–89

dental x-rays, proper setting for, 153–54, 159–60

Detroit Edison nuclear accident, 89

deuterium (heavy hydrogen), 6

developmental effects of ionizing radiation, 109–11; birth defects, 109; diagnostic radiologic examinations, 110–11; stage of fetal development, 109–10. *See also* ionizing radiation

Diablo Canyon nuclear reactor, earthquakes and, 92–93

diagnostic x-rays, 138

dilution/dispersion method of nuclear waste disposal, 99

diseases, radiopharmaceuticals for, 144

DNA: microwave radiation and, 128–29; mutations and, 112–13; sunbathing and ultraviolet light effect on, 130–31

doctors, x-ray abuse and, 157

dosages of radiation: age, as variable in, 52–53; average human, lifetime, 47–48; biological injury and dosage levels, 108–109, 110; birth defects and, 111; bone marrow absorption, 49–51; breast cancer and, 121; exposure variables and, 50; geography and, 40–41; "heaviness of radiation" and, 104; Hiroshima and Nagasaki survivors, information about, 107, 109, 118, 119; individual dose threshold, cancer and, 119; levels of, 108–12, 114; linear non-threshold dose-response hypothesis, 122; in medical treatment, 48–49; microwave, 125–26, 128; pregnancy and, 110–11; from radiopharmaceuticals, 50–54; rad/rem correlations and, 32, 108–10; skin cancer and, 120; speed of administering and, 105, 107; tissue damage and, 122; uranium mining sites, trailings at, 46–47; "whole-body," 38, 54

Down's syndrome, 110, 113, 127

drinking water, radon levels in, 44

education, radiation safety through, 176–77

effluents of nuclear fuel cycle, as distinct from wastes, 93

Einstein, Albert, 22

electrical activity, 11. See also electromagnetic radiation

electricity, 4–5; glass rods, rubbing, 5; ions and, 10; nuclear reactors producing, 82, 197. See also charges

electromagnetic radiation(s) (energy), 11, 13, 14–19; energy shifts of, 17–18; as forms of light, 13–14, 15–16; gamma radiation, 17, 18; heat and, 14–15; human tissues and, 16, 18; infrared radiation, 14; ionizing radiations, 18–19; microwave frequencies, 14; nonionizing radiations, 18; quanta, 13; radio-frequency radiations, 14; red light, 15; spectrum of (chart), 15; ultraviolet radiation, 16, 18; wavelike pulsation of, 13; x-ray, 16–17, 18

electromagnetic radiation(s) (energy), products and operations emitting, 54–59; lamps, 55–56; microwave devices, 56–59. See also location of radiation; radioactivity

electromagnetism, 7; interference of cardiac pacemakers and, 63

electrons of atom, 4, 5, 6; antielectron, 8; movements of, as energy shifts, 17; shells of, 9; volts, 28

electrostatic generator, 28

elements, stable, radioactive isotopes of, 25

emanations from degenerating atomic nuclei. See radioactivity

emergency core-cooling system (ECCS), nuclear reactors and, 91

Energy, Department of, 83, 188

energy: linear energy transfer (LET), 32; as mass of atoms, 4; microwave, absorption of, 57; production of, 22; shifts in atoms, 17–18; solar, 74. See also electromagnetic radiation

Environmental Protection Agency (EPA), 86, 94, 99, 190

equator, as effect on cosmic rays, 37–38

equipment, faulty, unnecessary x-rays and, 155–56

excitation, energy as, 18

exposure, radiation, stages of, 106–107
extra-high-frequency waves (EHF), 14
eyes: infrared light and damage to, 132; lasers, benefits and damage by, 134, 136, 144–45; microwave radiation and damage to, 126, 128; ultraviolet light and, 16, 130. *See also* body, human

fallout, nuclear, 65
Federal Communications Commission, 192
Federal Power Commission, 75
fetal development: ionizing radiation and, 109–10; nonionizing radiations and, 125
fission, in nuclear fuel cycle, 78–80; bombs, 78, 79; chain reaction, 78; critical mass, 78, 80; vs. fusion, 79
flare activity of sun, cosmic rays and, 36
fluoroscope, 138, 158–59
Food and Drug Administration (FDA), 56, 60, 135, 155, 156, 165, 166, 167, 169–71, 185–87
food(s): internal radiation and, 41–43; preservation, radiation for, 148–49
"footprints" of radiation, 30
Fort St. Vrain, Colorado, nuclear accident at, 90
fossil fuels, 73–74
Franklin, Benjamin, 4–5
frequency: of electromagnetic radiation, 13; microwave, 14; radio-, radiation as, 14; waves, types of, 14
fruits, radioactivity levels in, 43
fuel cycle, nuclear, 75–85; breeder reactors, 85; fission, 78–80; heavy-water reactors, 85; high-temperature gas reactor, 84; light-water reactor, 83–84; nuclear reactor, 80–83; rods, 77–78; spent fuel, as nuclear waste, problem of, 94–96;

uranium, 75–78. *See also* nuclear energy
fusion vs. fission process of energy generation, 80; hydrogen and, 79

gamma rays, 12–13, 17, 18, 40, 41, 143, 150
gases: natural, 45; radioactive, earth's formation by, 39–40
gas reactor, nuclear, high-temperature, 84
Geiger counter, 30
Geiger, Hans, 30
generator, electrostatic, 28
genes, 112; effects of ionizing radiation on, 112–13; sunbathing and ultraviolet light effects on, 130–31
geography, cosmic ray exposure and, 37
G.I. series, radiation doses and, 50
glass rods, rubbing, for electricity, 5
government. *See* nuclear regulatory establishment
graphite, x-ray safety and, 177–78
gravity, atomic theory and, 7

half-life, in radioactive decay, 23
Hanford nuclear accident (Richland, Washington), 90
Health, Education and Welfare, Department of, 49, 54, 155
Health and Human Services, Department of, 71, 183
health-related radiation exposure, 47–54; cancer treatment, 48; radiopharmaceuticals, 50–54; x-rays, 49–50, 51. *See also* body, human; tissue, human
heat, atomic generation of, 14–15
"heaviness" of radiation, body damage and, 104
heavy-water reactors, 84–85
helium, 6, 9
heredity, chemicals and, 112–13

high-temperature gas nuclear reactor, 84

Hiroshima and Nagasaki bombs, knowledge about radiation from, 107–108, 109, 110, 113, 118, 119

hormonal imbalance, microwave radiation and, 128

housing construction, radioactivity in materials in, 41, 46

hydroelectric plants, 74

hydrogen, 9; atoms of, 6; in fusion process, 79; -3, 45

hyperons, 8

hyperthyroidism, 53–54, 66. See also thyroid

individual dose threshold of radiation, cancer and, 119

industrial uses of radiation, 150

infrared light: in medical treatment, 145; safety precautions for, 167

infrared radiation, 14, 131–32

insect control, radiation and, 150

Institute for Nuclear Power Operations, 162

interaction, strong and weak, between atoms, 7

internal radiation, 41–44; calcium, 43; cigarette smoking as source of, 46; food and water, 41–42; phytoplankton, 44; potassium-40 and, 43; radon, 44

International Dumping Treaty, on nuclear wastes, 99

iodine-131: hyperthyroidism treated with, 53–54; nuclear testing and, 66; as waste, nuclear, 86

ionization chamber (radiation chamber), 30

ionizing radiations, 18–19; types of (chart), 19

ionizing radiations, making safer, 153–63; consumer products, 160–63; dental setting, 159–60; medical and dental x-rays, 153–54; medical setting, proper, 158–59;

nuclear power industry, 153; x-rays, unnecessary, 154–58

ionizing radiations, products emitting, 59–67; cardiac pacemaker, 63; lasers, 60–62; nuclear testing, 65–67; smoke detectors, 59; televisions, 62; uranium coloring, 59–60; video display terminal (VDT), 62–64; watches, 59. See also location of radiation; mutation and cancer; nonionizing radiations

ionosphere, as protection against radiation, 38

ions, 10–11

iron-55, nuclear testing and, 66

isotopes, 6; radioactive, of stable elements, 25

Kennedy, Edward M., 103

Klaproth, Martin Heinrich, 75

krypton-85: nuclear testing and, 66; as nuclear waste product, 86

Labor, Department of, 188

lambda, as hyperon variation, 8

lamps, as source of electromagnetic radiation, 55–56

land burial, shallow, of nuclear wastes, 100–101

lasers: eye damage and, 134, 136; as health hazards, 133–34; as ionizing radiation emitters, 60–62; in medical treatments, 144–45; in radiation therapy, 144–45; safety standards for, 135–36, 167

latent period of radiation exposure, 106

Lawrence Livermore Laboratory nuclear accident, 90

lead, 20, 21, 22; -206, 21; as x-ray shield, 17

leakage of nuclear wastes, 97, 98

legal protection of job, x-rays and, 154

leptons, 8

LET. *See* linear energy transfer

leukemia, 18, 19; Hiroshima and Nagasaki survivors and, 107, 109, 118, 119; in St. George, Utah, 103–104

licensing of x-ray technicians, 176

light, visible: electromagnetic radiation, as form of, 13–14, 15–16; in medical treatment, 145; as possible hazard, 133–34

light-water reactor, 83–84; ecological dangers of, 84

linear accelerator (linac), 28

linear energy transfer (LET), 32, 104

linear nonthreshold dose-response hypothesis, 122

liquids, radioactive, as nuclear wastes, problem of, 96–97

lithium, 6; -7, as first artificially produced nuclear reaction, 27

location of radiation, 35–72; electromagnetic radiation, products and operations emitting, 54–59; health-related human contributions to, 47–54; ionizing radiation, products emitting, 59–67; natural radiation, 35–47; personal radiation inventory, 67–72

lung cancer, 120. *See also* cancer

magnetism, 11. *See also* electricity

magnetopause, as protection against cosmic rays, 38

mammography, 139–40, 145, 158; cancer and, 120–22

mass: critical, in fission bombs, 78, 80; energy and, 4

median lethal dose of radiation, 108

medical diagnosis, uses of radiation and, 137–42; CAT scan, 140–42; x-rays, 137–40

medical treatment, use of radiation in, 142–45; infrared light, 145; laser, 144–45; radiopharmaceuticals, 144; ultraviolet rays, 145; visible light, 145; x-ray therapy, 142–43

medical x-rays, proper setting for, 158–59

medicine, defensive, unnecessary x-rays and, 109

melanoma, microwave treatment for, 143

"meltdown," nuclear, 88; near, at Three Mile Island, xi–xiii, 88–91

mercury vapor lamps, dangers of, 166–67

mesons, pi and mu, 7–8

meteorology, radiation and, 149

Mev (million electron volts), 28, 29

microwave frequencies, 14

microwave ovens, safety precautions for, 168–71

microwaves, as source of electromagnetic radiation, 56–59, 124–30; animal studies, 124–26; cancer, 128–30; as danger to cardiac pacemakers, 63; humans, effects in, 126–28; melanoma treated with, 143; ovens, 56–59; radar, 56; safe levels of body absorption, 124; satellites, 57

military applications of radiation, 147–48

milk, radioactivity in, 43

Millstone One nuclear accident, 91

mining: radiation from, 45, 46, 47; solution, 76; strip (or pit), 76

moderators, in nuclear reactors, 80

molecular motion, 10

molecules: chemical, 10; colliding, 11

mu-meson (muons), 7–8

mutagen, 112

mutation, as effect of ionizing radiation, 103–14; acute radiation sickness, 107–109; biological damage, 104–105; cause and effect relationships, 104; cells, body, reaction to radiation, 106;

developmental effects, 109–11; genetic effects, 112–13; Hiroshima and Nagasaki atomic bombs and, 107–13, 118, 119; as late effect of radiation exposure, 113–14; stages of radiation exposure, 106–107; in St. George, Utah, 103–104. *See also* cancer

Nagasaki. *See* Hiroshima
National Bureau of Standards, 130
National Cancer Institute (NCI), 131, 139, 140, 187
National Council on Radiation Protection and Measurement, 173, 174
National Institute for Occupational Safety and Health (NIOSH), 135, 187, 188
National Institutes of Health (NIH), 187
natural radiation, 35–47; cosmic rays, 35–39; internal radiation, 41–44; technologically enhanced, 44–47; terrestrial sources of, 39–41. *See also* location of radiation
neutrino, 8
neutrons, 6, 7, 12; as stabilizers of protons, 20
nonionizing radiations, hazards of, 18, 123–36; chemistry of, 123; infrared radiation, 131–33; microwave and radar radiation, dangers of, 124–30; protective standards, 134–36; ultraviolet light, 130–31; visible light, 133–34
nonionizing radiations, reducing risks in using, 164–71; infrared light, 168; laser beams, 167; microwave ovens, 168–71; ultraviolet light, 164–67. *See also* ionizing radiations
nuclear energy, as positive benefit of radiation, 145–47, 197. *See also* energy; fuel cycle, nuclear

"nuclear garbage," 84. *See also* wastes
nuclear industry, 73–102; accidents, threat of, 85–86, 89–90, 91; capacity of, in U.S., 74–75; fossil fuels and, 73–74; fuel cycle, 75–85; hydroelectric plants and, 74; plant safety, 87–93; radioactive materials, transportation of, 86–87; security of plants, 93; solar energy and, 74; testing, nuclear, as ionizing radiation emitter, 65–67; wastes, disposal of, 93–102
nuclear medicine, 144
nuclear reactors, reducing risks of, 161–63; emergency core-cooling system (ECCS) and, 91
Nuclear Regulatory Commission (NRC), 70, 93, 94, 162, 190–91, 193
nuclear regulatory establishment, 183–95; Center for Disease Control, 187; Commerce, Department of, 189; Consumer Product Safety Commission (CPSC), 191–92; Defense, Department of, 188; Environmental Protection Agency (EPA), 190; Federal Communications Commission (FCC), 192; Food and Drug Administration (FDA), 185–87; Health and Human Services, Department of, 183; Labor, Department of, 188; National Institutes of Health (NIH), 187; Nuclear Regulatory Commission (NRC), 190–91, 193; standards, setting of, 194–95; state agencies, 193–94; Transportation, Department of, 189; United States Government Radiation Protection Responsibilities (chart), 184–85; Veterans Administration (VA), 192
nuclear safety analysis center (NSAC), Three Mile Island investigation and, 162

nucleus, atomic, 4–8; neutrons, 6, 7; probing with "subatomic" particle "gun," 26; weight of, 6

oceans, burial of nuclear wastes in, 98, 99
Occupational Safety and Health Administration (OSHA), 188
Office of Nuclear Material Safety and Safeguards, 63
Office of Public Awareness (of EPA), 190
Office of Radiation Programs (of EPA), 190
oligospermia, 133
omega-minus particle, 8
operations emitting radiation. See electromagnetic radiation
ores, radiation from, 45
organs: critical, radiopharmaceuticals and, 52; reproductive, male, dangers of x-rays to, 157–58
oxygen, 10

pacemakers, cardiac, as ionizing radiation emitters, 63–64; dangers to, 64–65; electromagnetic interference and, 62–63
particle accelerators, artificial production of radioactivity by, 27–28
particles, subatomic, 25–27; omega-minus, 8
patients: ownership of x-ray films by, 173–74; physical condition of, radiation exposure and, 52
personal radiation inventory, 67–72; information, obtaining, 71–72; ionizing radiation on "whole-body," estimating (chart), 68. See also location of radiation
photon(s), 8, 14, 15–19; electron movements and, as energy shifts, 17; high-energy, 18
physics, laws of, 6–7
phytoplankton, internal radiation and, 44

pi-meson (pions), 7–8
pitchblende, 20–21, 75
pit mining, 76
plant safety and security, nuclear, 87–93; deaths, 88, 90, 91; Diablo Canyon and, 92–93; "meltdown," 88; Millstone One accident, 91; near-calamities, 89–90, 91; Three Mile Island accident, primary cause of, 90–92
plants, hydroelectric, 74
plutonium: nuclear waste and, 86; -238, 63, 64; -239, 96
poles, earth's, cosmic rays and, 37, 38
polonium, half-life of, 23
populations, particular, cancers linked to radiation in (chart), 115–16, 119
positron (antielectron), 8
potassium, 25, 43, 45
pregnancy: ionizing radiation and, 107–11; x-rays and, 157, 173, 174
primary radiation, as cosmic ray, 35
prodrome phase, of acute radiation sickness, 107–108
products emitting radiation. See electromagnetic radiation; ionizing radiation; nonionizing radiation
Project Pandora, microwave cancer and, 129
protactinium-234, 21
protons of nucleus, 5–6, 7, 12; antiproton, 8; artificial production of radioactivity and, 26–27; neutrons and stability of, 20
purification of cooling fluid, in nuclear reactor, 84

"quality assurance," x-rays and, 178
quanta, 13
quarks, particles and, families of, 8

rad (radiation absorbed dose), 32; rem/rad absorption correlation, 104–105, 108–10

radar: systems, electromagnetic radiation from, 56; use of radiation in, 149. *See also* microwaves

radioactive materials, transportation of, 86–87

radioactivity, 12–14; alpha particles, 12; beta particles, 12; as degeneration of atom, 12; gamma rays, 12–13, 17, 18, 27, 40, 41, 43, 143, 150

radioactivity, origin and measurement of, 20–32; artificially produced radioactivity, 25–29; decay, radioactive, 20–25; proton-neutron stability quotient, 20; tracking and quantifying radiation, 29–32. *See also* electromagnetic radiation(s)

radio-frequency radiations, 14

radioisotope scan, tumors and, 159

radionuclides, 50–54, 144

radiopharmaceuticals, as health-related radiation exposure, 50–54; critical organs and, 52; radiation therapy and, 144; sodium in blood and, 51; target organ (e.g., thyroid gland), 52; thyroid treatments with, 53; variables in use of, 52–53

radium, 59; series (chart), 24; -226, 21

radon, internal radiation and, 44, 45, 47

"rare earth," screens, 177

rays, gamma, 12–13, 17, 18. *See also* cosmic rays

reactors, nuclear, 80–83; breeder, 85; heavy-water, 84–85; high-temperature gas, 84; light-water, 83–84

recovery period, of radiation exposure, 104

red light, as electromagnetic radiation effect, 15, 16

referral criteria, reducing radiation exposure by, 174–75

reflection of microwave energy, 57

regulatory agencies, nuclear. *See* nuclear regulatory establishment; state regulatory agencies

relative biological effectiveness (RBE), 32

rem (radiation equivalent man), 32. *See also* rad

reproductive organs, male, dangers of x-rays and, 157–58

rocks, radioactivity and, 40

rods: cadmium-plated, 80; control, in nuclear reactor, 80; electricity and glass and sealing-wax rods and, 5; in nuclear fuel cycle, 77–78; zinc and copper, 11

roentgen unit, 31

Roentgen, Wilhelm K., 31, 105

rubidinum-87, 40, 86

safety of radiation, innovations in, 172–79; ALARA safety standard, 195; computer contributions, 175–76; level of exposure per x-ray, reducing, 177–78; professionals' responsibility, 172; "quality assurance," 178; training, licensing and education, 176–77; x-ray exposure, reducing number of, 173–75

St. George, Utah, leukemia levels at, 103–104

salt beds, for disposal of nuclear wastes, 98

satellites, electromagnetic radiation from, 57

scattering process, of cosmic rays, 37

Schwan, Dr. Herman, 124

scintillator, 30–31

screens, radiation. *See* "rare earth" screens

sea disposal of nuclear wastes, 99

seafood, radioactivity in, 44, 46

sealing-wax rods/glass rods, electricity and, 5

secondary radiation, as cosmic ray, 35

shield(s): earth's protective, from radiation, 36–38; lead, for x-rays, 17; in nuclear reactor, 81–82

sickness, acute radiation. *See* acute radiation sickness

sigma, as hyperon variation, 8

skin: cancer, 120, 130–31, 132, 145, 164–65; infrared light and damage to, 132; ultraviolet light and damage to, 130–31. *See also* cancer

smoke detectors, as ionizing radiation emitters, 59

sodium: in blood, radiopharmaceuticals and, 51; ions, 10

soil build-up, radioactivity and, 40, 43

solar energy, 74

somatic effects, acute, radiation sickness and, 107

spent fuel, problem of, 95–96

standards, setting of, for nuclear industry, 194–95

stars, cosmic rays from, 36

state regulatory agencies, 193–94

Stoessel, Ambassador, 129

stomach cancer, 120. *See also* cancer

strategic arms proliferation, as risk of nuclear power, 197. *See also* bombs

strip mining, 76

strontium-90, 43, 66, 86, 96, 101

subatomic particles, 5, 7–8; division of, 8; as "gun" for probing nucleus of atom, 26; mesons, pi and mu, 7–8; omega-minus, 8; radioactivity, as degeneration of, 12

sunbathing, hazard of, 130–31; skin cancer and, 164–65

sun, cosmic rays from, 36

sunlamp injury, 165–66

super-high-frequency waves (SHF), 14

synchrotons, 29

tanning booths, dangers of, 166

target organ, radiopharmaceuticals and, 52

technetium-99, 144

technology, enhancing of natural radiation by, 44–47

teeth. *See* bone(s)

televisions, as ionizing radiation emitters, 62, 161

Tennessee Valley Authority Brown's Ferry nuclear accident, 89–90

terrestrial sources of radiation, 39–41; body exposure, external, from natural radioactivity (chart), 42

testing, nuclear, 65–67

therapy, radiation. *See* medical treatment

thermography, 145

thermonuclear weapons, 79, 147

thimble chamber, 31

Thompson, Llewellyn, 129

thorium, 40; -230, 21; -232, 85; -234, 21

Three Mile Island, nuclear accident at, issues involving, ix–xi, 87, 88, 137, 146, 153, 162; emergency core-cooling system shutdown at, 91; faulty designs of reactor, 92; near "meltdown" at, xi–xiii, 88–91; nuclear safety analysis center (NSAC) investigation, 162; primary cause of accident, 91

thyroid: cancer of, 119; hyperthyroidism, 53–54, 66; radiopharmaceutical treatment for, 53

tissue, human: composition of, radiation and, 16, 18; infrared radiation and, 133; linear nonthreshold dose-response hypothesis of ionizing radiation and, 122; nonionizing radiation and, 123; sensitivity to radiation, 106. *See also* body, human

"tracers," radionuclides as, 50–51

tracking and quantifying radiation, 29–32; ionization chamber (radiation chamber), 30; linear energy transfer (LET), 32; rad (radiation absorbed dose), 32; relative biological effectiveness (RBE), 32; rem (radiation equivalent man), 32; roentgen unit, 31; scintillator, 30–31; thimble chamber, 31

transmission of microwave energy, 57, 58–59

Transportation, Department of, 189

transportation of nuclear material, 197

transuranic elements, in nuclear wastes, 95

tritium, 45, 59, 66, 86, 99, 101

tuberculosis, x-ray abuse and, 156, 185

tumors, radioisotope scan and, 159

tungsten halide bulbs, dangers of, 166–67

ultra-high-frequency waves (UHF), 14

ultraviolet light: avoiding exposure to, 164–67; DNA, effects on, from sunbathing, 130–31; emitting standards, 135; possible damage from, 130–31; radiation, 16, 18; rays, in medical treatment, 145

units of measurement, in radioactivity, 31–32

uranium, 20–21, 75–76, 85; as by-product, 76; coloring, as ionizing radiation emitters, 59–60; mills in U.S. (1976), 77; mining, 45–47, 71, 76–77; in nuclear fuel cycle, 75–78; as nuclear waste, 86; (radium) series (chart), 24; -234, 21; -235, 77–80, 84; -238, 20–21, 40, 44, 84, 85

Van Allen belts, 36, 45

vapor lamps, mercury, dangers of, 166–67

Veterans Administration (VA), 192

video display terminal (VDT), as ionizing radiation emitters, 62–63, 64

virus theory of cancer, 117

visible light. *See* light, visible

voltage multiplier, 28, 29

V-particles, 8

wastes, nuclear, disposal of, 93–102, 197; classification of wastes and corresponding problems, 94–96; and effluents, difference between, 93; location, present, of burial sites, 100–101; methods of disposal, various, 98–99; quantity of, 97–98; risks and responsibilities, 102; transuranic elements in, 95

watches and clocks, as ionizing radiation emitters, 59

water, internal radiation and, 41, 44; running, action of, 40

wave lengths, 13–16; light, types of, 15–16; microwave frequencies and, 14

"whole-body" doses of radiation, 38

workers, radiation, federal standards for (e.g., aircrew members), 37, 46, 54

World Health Organization, 129

xi, as hyperon variation, 8

x-ray(s), 16–17, 18; bone marrow exposure to (chart), 51; cancer, as cause of, 118; CAT scans, 140–42; dental, 153–54, 159–60, 172; diagnostic, 138; doctors, abuse by, 157; exposures to, reducing number of, 173–75; first case of cancer and, 105; fluoroscope, 138, 158–59; as health-related radiation exposure, 49–50, 51; lead, as shield for, 17; level of exposure per, reducing, 177–78; job protection and, 154; machine, for artificially producing radioactivity, 29; mammography, 139, 158; medical, 137–40, 142–43, 153, 158–59, 172; organs, male reproductive, dangers to, 157–58; ownership of films, patient or doctor, 174; pregnancy and, 110–11, 157; "quality assurance" of, 178; technicians, licensing of, 176; therapy, 142–43; tuberculosis, as abuse of, 185; unnecessary, 154–58. *See also* CAT scanners; radioisotope scan

x-ray technicians, licensing of, 176

zinc ions, 11

ABOUT THE AUTHORS

MARTIN ECKER received his B.A. from Cornell University, attended New York University School of Medicine and completed a medical internship at Beth Israel Medical Center. He did his residency in diagnostic radiology at the combined programs at North Shore University and the Cornell University–New York Hospitals, served as chief resident radiologist at the North Shore Hospital and was formerly a professor of diagnostic radiology at the Yale University School of Medicine. Presently he is the director of radiology at the White Plains Medical Center in White Plains, New York. He resides with his wife and three children in Stamford, Connecticut.

NORTON J. BRAMESCO, a science writer, is creative director of the Healthmark Division of Medcom, an organization that specializes in communication to the medical profession and to lay audiences on medical subjects. He is also a highly regarded writer of medical advertisements and has published popular articles on medical and nonmedical subjects. He has published as well mystery fiction under the name of Bram Norton. He is married and lives in New York City.